Elements of Chemical Thermodynamics

Second Edition

Leonard K. Nash
Harvard University

DOVER PUBLICATIONS, INC.
Mineola, New York

Copyright

Bibliographical Note

This Dover edition, first published in 2005, is an unabridged republication of the second edition, published by Addison-Wesley Publishing Company, Inc., Reading, Massachusetts, 1970.

Library of Congress Cataloging-in-Publication Data

Nash, Leonard Kollender, 1918–
Elements of chemical thermodynamics / Leonard K. Nash.
p. cm.
Originally published: 2nd ed. Reading, Mass. : Addison-Wesley Pub. Co., 1970, in series: Addison-Wesley series in the principles of chemistry.
Includes bibliographical references and index.
ISBN 0-486-44612-3 (pbk.)
1. Thermodynamics. I. Title.

QD504.N37 2005
541'.369—dc22

2005048408

Manufactured in the United States of America
Dover Publications, Inc., 31 East 2nd Street, Mineola, N.Y. 11501

Preface

Let no man say I have done nothing new;
the arrangement of the material is new.
B. PASCAL

From purely thermal data, to calculate the position of equilibrium in a chemical reaction perhaps not even yet achieved: *that* is the capacity of thermodynamics which is at once most striking and most important to the beginning chemist. The entire train of argument in this text has been shaped to bring that calculation within reach, and I have treated no topic lying away from the path that leads to development and application of the basic thermodynamic criteria for the condition of equilibrium.

In this second edition, as in the first, I have everywhere sought to highlight the physical content of thermodynamics, as distinct from its purely mathematical machinery. Because elaboration of this analytical machinery may obstruct rather than forward the beginner's efforts to grasp the essential physical ideas, I have used only a dozen of the most elementary operations of the calculus—all of which are explained in Appendix II. In several developments of the argument, sophisticated elegance must then give way to a simple clumsiness, and certain other developments (e.g., those demanding explicit use of partial derivatives) cannot even be attempted. In my view, however, these losses are wholly outweighed by the possibility of that gain I have pursued to the exclusion of all others: the focus of attention remains forever centered on the physical concepts involved. This same concern has motivated an effort to provide the reader with several indications of the physical correlates of the entropy concept. That is, while stressing repeatedly that we transact the business of classical thermodynamics in complete independence of all hypotheses about the nature of matter, energy, and entropy, I have nevertheless seized every opportunity to give rough statistical *interpretations* of what it is we obtain from entropy *calculations*.

Though this second edition is some 50% longer than the first, the increase does *not* reflect any great enlargement of coverage. Instead, with a view to facilitating the beginner's growth to mastery of what remains essentially the same body of material, I have greatly enlarged the explanatory discussions of the nature and use of the fundamental thermodynamic concepts. Most notably, the first chapter now offers a very much expanded account of the concepts of heat, work, and reversibility. And the last chapter offers an expanded account of the application of the free-energy concept to ideal solutions, and bases on this account a more fundamental analysis of the colligative properties of such solutions. The number of problems (collected in Appendix III) has undergone an almost threefold expansion, and thirty fully worked illustrative examples have been dispersed throughout the text. The only other significant addition is the provision (in Appendix I) of some 80-odd references to books and articles that treat various aspects of thermodynamics in a manner likely to be enlightening to the beginner. This bibliographical material abstracts, and extends to 1969, the more comprehensive listing given in the author's "Resource Paper on Elementary Chemical Thermodynamics," *J. Chem. Educ.* **42**, 64 (1965).

This text assumes no more than a sound background in high-school mathematics and physics, and familiarity with the leading quantitative concepts of the traditional introductory college chemistry course. Though the problems now constitute a fairly comprehensive set, and though many of the individual items are quite searching, all but a few of the problems still require no use of the calculus, and even the few demand use of only the very elementary operations developed in Appendix II. Now certainly this first introduction to thermodynamics cannot stand in lieu of a more advanced treatment of the same subject, but it should permit that advanced course to go faster and further, while at the same time providing more challenging intellectual fare for the beginning student. Thus, for example, a fully mathematized axiomatic treatment of thermodynamics in the style of Gibbs is a rewarding intellectual experience for the advanced student, while to grasp for the first time the *full* significance of the simple Carnot cycle is also a thoroughly rewarding intellectual experience readily accessible to the beginner.

Amongst the myriad equations in the body of the text, the most important relations have been systematically distinguished by assigning numbers to them; the letter assignments, on the other hand, serve only to facilitate identification of a relation operative in the immediately succeeding text, or in a problem. The symbol for the Gibbs free energy has been changed from F to G, thereby conforming to what seems to have become common practice. Adoption of the officially recommended sign convention for work terms still being far from common practice, I have retained the older sign convention with which I feel more at ease.

For several important lines of argument I have drawn on C. N. Hinshelwood's *Structure of Physical Chemistry* (Oxford University Press, 1951) and, still more, on E. F. Caldin's *Introduction to Chemical Thermo-*

dynamics (Oxford University Press, 1958). With the permission of its publisher, I have also reproduced from this last work one short quotation of text (on p. 91) and the two tables I have numbered 5 and 6. Figure 35 is redrawn from Hildebrand and Scott's *Solubility of Non-Electrolytes* (Reinhold, 1950) with the consent of the publisher. Figure 32 and the immediately accompanying text were suggested by Jürg Waser's *Basic Chemical Thermodynamics* (Benjamin, 1966). Four problems, drawn in substance from other texts, carry the names of the authors to whom I am indebted.

For their willingness to read the manuscript of this work, and for many constructive suggestions arising from those readings, I am most grateful to Francis T. Bonner, the editor of this series, and to Jerry A. Bell, Donald D. Fitts, Irving M. Klotz, and Mark W. Zemansky. Miss Martha Stark lent assistance with debugging of the manuscript; she and George Eliopoulos gave valuable help with proofreading; and he and Preston Black separately reviewed all the new problems and worked most of them. My dear wife has as always exercised her talents not only in endless proofreading but also in the care, feeding, and toleration of a preoccupied author—who is of course solely responsible for any residual defects in this work.

Cambridge, Massachusetts L. K. N.
August 1969

Contents

1

Introduction

Thermodynamics is a notably abstract science with innumerable concrete applications. It unites extreme generality with extreme reliability to a degree unsurpassed by any other science. The magnificent generality of thermodynamics arises from the abstractness of its fundamental concepts: stripped insofar as possible of everything referring to specific things and events, such concepts are fitted for the interpretation of relations of the most diverse varieties. The extraordinary reliability of thermodynamics arises because, while calling on only a minimal array of axiomatic postulates, it cunningly contrives to discuss material phenomena without making any assumption whatever about the constitution (atomic or otherwise) of matter.

Thermodynamics treats of *systems*—parts of the world that definite boundaries separate conceptually (and often, with a good degree of approximation, physically) from the rest of the world. The condition or *state* of such a system is regarded as thermodynamically defined as soon as values have been established for a small set of measurable parameters.[1] These parameters are so chosen that any particular state of the system will be fully and accurately reproduced whenever the defining parameters take on the set of values descriptive of that state. Temperature, pressure, volume, and expressions of concentration (e.g., mole fractions) are the parameters most used by chemists, and temperature in particular is distinctive of thermodynamic analyses generally. But not every state of every system can be characterized by a single well-defined temperature (and pressure), and from this circumstance arises two major restrictions on the applicability of classical thermodynamics. First, given the (Maxwellian) distribution of molecular velocities, a single molecule, or even a small group of molecules, does not have a definite temperature. Only to *macroscopic* systems will a temperature be assignable, and only to such systems will thermodynamics be applicable. Second, even a macroscopic system will manifest local inhomogeneities of temperature (and pressure) while it is undergoing rapid change. An entire macroscopic system will be

---

[1] For this and all other bibliographical notes indicated by consecutive superscript numbers, see Appendix I on p. 140.

characterizable by a unique temperature (and pressure) only when that system stands in an unchanging state of *equilibrium*—or in a quasi-static state only infinitesimally different from a true equilibrium state. And it is with such states alone that classical thermodynamics concerns itself.

Focused on the equilibrium states of macroscopic systems, classical thermodynamics tells us *nothing about the paths* by which different states may be connected, and *nothing about the rates* at which the paths may be traversed and the states attained. Thermodynamics is then, if you please, rather more the science of the possible in principle than the science of the attainable in practice. What thermodynamics finds impossible we cannot hope to achieve, and we are spared the investment of effort in a vain endeavor. But to achieve in practice what thermodynamics finds possible in principle may still require an immense endeavor. For example, thermodynamic calculations show that, under pressures of 30,000–100,000 atmospheres, diamond should be formed from graphite at temperatures of 1000–3000°K. However attempts to achieve this conversion were for long totally unsuccessful. Having confidence in thermodynamics, one attributes these failures to slowness of the reaction, and one perseveres in the endeavor to achieve in practice what thermodynamics indicates as possible in principle. And in 1954 the synthesis of diamond was at last achieved—through the discovery of an effective catalyst, and the construction of equipment competent to maintain the required conditions for hours rather than seconds.[2]

TEMPERATURE

The theoretical term "temperature" first acquires empirical relevance only when we learn how temperature is to be determined. Let it now be specified that temperature is to be measured with a constant-volume gas thermometer, shown schematically in Fig. 1. When the gas in the sealed bulb has come to equilibrium at the temperature (T) prevailing in the enclosure, by appropriate manipulation of the leveling bulb, we bring the mercury meniscus always to the indicated fiducial mark. Having thus assured constancy of the gas volume (V), we obtain the pressure (P) by measurement of the height h. To establish what we mean by temperature, we have then only to write, for the measurements made at any two different temperatures:

$$T_2/T_1 \equiv [P_2/P_1]_V,$$

where the subscript V underlines the constant-volume requirement. The temperature ratios determined from this formula prove unhappily variable with the identity and pressure of the gas present in the bulb. But whatever the gas used, this variability disappears when we repeatedly determine the pressure ratio with progressively diminished amounts of gas in the bulb,

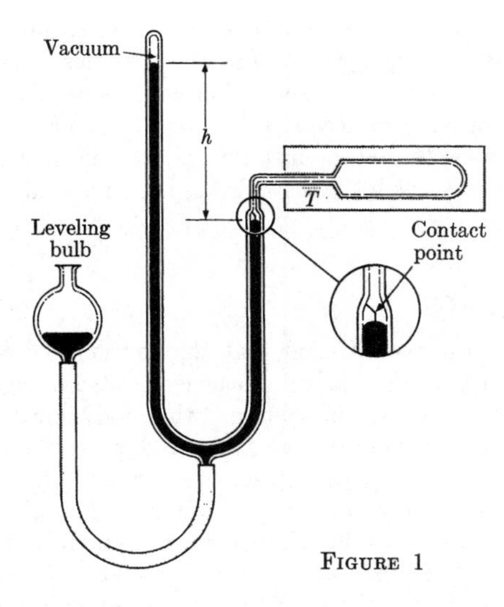

FIGURE 1

and then extrapolate to the limit $P \to 0$. That is, with *any* pure gaseous species we obtain the *same* temperature ratio from the limiting pressure ratio:

$$T_2/T_1 \equiv \lim_{P \to 0} [P_2/P_1]_V.$$

To pass from the temperature *ratios* so defined to numerical values for individual temperatures, we need only adopt a convention that assigns some particular numerical magnitude to the temperature of some one standard reference point. By international agreement,[3] we assign 273.16 as the temperature at the "triple point" at which ice, water, and water vapor coexist in equilibrium.* Letting the subscript tp signify the triple point, and dropping all other subscripts, we find that

$$T \equiv 273.16 \left\{ \lim_{P \to 0} [P/P_{tp}]_V \right\}.$$

This equation expresses the so-called "ideal-gas temperature scale" and, anticipating a conclusion reached on p. 67, we shall henceforth distinguish readings on this scale as °K, e.g., 273.16°K.

---

* As a reference temperature the triple point is preferred to the rather less precisely fixed normal melting point of ice in contact with air-saturated water. The figure 273.16 is chosen with a view to maintaining the familiar Celsius-scale interval of 100° between the normal freezing point and boiling point of water: 273.15 and 373.15 respectively on the scale we have here constructed. (See problem 1.)

In convenience, other thermometric devices far surpass the constant-volume gas thermometer. And in the extremes of very high and very low temperatures, the use of these other devices becomes a matter not merely of convenience but of necessity.[4] Ultimately, however, all such alternative devices are calibrated against (or at least indirectly referred to) the gas thermometer—which thus establishes the empirical meaning of the concept "temperature."

HEAT AND WORK

In much the same period that the equivalence of heat and work was first established, the limited efficiency of steam engines posed a problem from which developed the science of thermodynamics. Today we no longer conceive heat in the original "thermo-dynamic" sense, as a caloric fluid that "flows." Yet that conception is commemorated in a familiar unit—the calorie—which represents approximately the quantity of heat required to warm 1 gm of water 1°K. Note however that the fundamental unit of both heat and work is the joule (= 1 newton-meter = 1 watt-second = 1 volt-coulomb), and the calorie is today a derived unit defined as 4.184 joules.[3] Because on the molar scale most chemical changes involve transfers of a *great* many calories, the unit of heat most used by chemists is neither the calorie nor the joule but the kilocalorie (= 1000 calories = 4184 joules).

Heat and work are different modes of energy transfer between a system and its surroundings. To indicate the *direction* of transfer, we use a sign convention that derives from the original thermodynamic concern with systems constituted by heat engines delivering a *work output* when supplied with a *heat input*. Thus work delivered *by* the system is called positive, and work supplied *to* the system is then appropriately negative—diminishing the *net* production of work. Heat supplied *to* the system is called positive, and heat given up *by* the system is then appropriately negative—since every heat loss means a smaller production of work.

Defining heat as energy transferred by thermal conduction and radiation, we may think to have distinguished it clearly from work, defined as energy transferred by other mechanisms. But such purely verbal definitions are wholly insufficient. We do better to proceed as we did with "temperature," by seeking for heat and work some simple operational definitions which, though incomplete, amply identify the empirical bearing of the abstract concepts.

For "heat" our definition takes as its cornerstone the so-called ice calorimeter shown schematically in Fig. 2. This conceptually simple but remarkably accurate device is primarily intended only for heat measurements at 0°C (273°K), but any of the manifold other ways of measuring heat can in principle be referred to the paradigmatic case constituted by the ice calorimeter.[5]

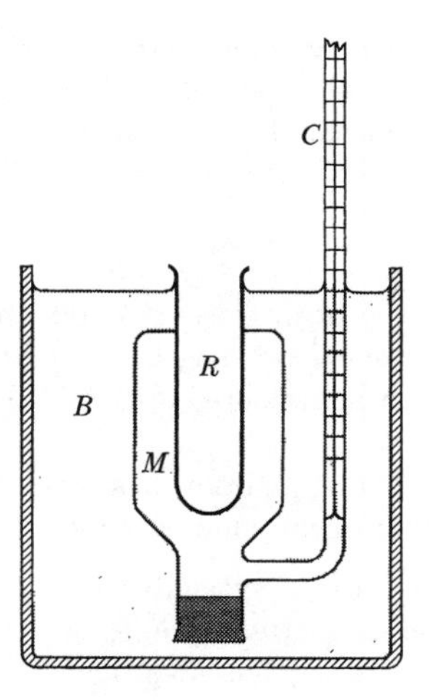

FIGURE 2

Imagine chamber M completely filled with a mixture of ice and water, and the stopper pressed home until the water meniscus stands within the graduated region of the capillary tube C. The same ice-water mixture is used to fill the outer bath B: by thus enforcing temperature equality, we ensure that there will be *no* heat transfer between M and the surrounding bath B. But if the temperature reigning in the inner chamber R deviates even minutely from 0°C, heat *will* be transferred through the rigid but thermally conductive wall between R and M. If, for example, we bring about in R an exothermic chemical reaction, the consequent transfer of heat will melt some part of the ice in the mixture that fills M. What will be the effect? The resultant mixture will of course differ not at all in temperature from the 0°C of the original mixture. However, due to the fact that at 0°C ice is substantially less dense than water (recall: ice floats on water), any melting of ice will be accompanied by a decrease in the volume of the mixture. Thus the transfer of heat from R to M will be reflected in a measurable subsidence of the water meniscus in C.

Now even as "pressure" is expressible in terms of cm Hg, just so a quantity of "heat" could easily be expressed in terms of the cm^3 of volume change produced by its transfer to or from an ice calorimeter. But actually we much prefer the joule and the calorie as units for heat, and a single additional determination suffices to convert all ice-calorimeter measurements to this basis. We have only to set up the ice calorimeter afresh, placing in R a small electrical resistor through which we pass a known current for a known time. We are then in a position to calculate the number of joules (or calories) of electrically generated heat transferred from R to M, and to read off from C the volume change produced by that transfer. And so we establish once and for all the conversion factor that permits us to pass from an observed volume change to a statement that some particular number of joules represents the "heat" transferred to or from the calorimeter.[6] The great simplicity of the basic calculation should be evident from the following three examples.

▶ *Example 1*

Into an ice calorimeter was inserted a 10.00 ohm resistance through which a current of 1.000 ampere was passed for 4.00 minutes. The consequent decrease in volume was measured as 0.653 ml. Determine the conversion factor:

$$\alpha = \frac{\text{Number of joules of heat transferred}}{\text{Number of ml volume change observed}}.$$

Solution. The formula for the electrically generated heat (in joules) is $\mathcal{I}^2\mathcal{R}t$, where $\mathcal{I}$ is the amperage, $\mathcal{R}$ the ohmage, and t the number of seconds during which the current passes. We have then:

$$\text{Heat} = (1.000)^2(10.00)(240) = 2400 \text{ joules.}$$

Therefore,

$$\alpha = \frac{2400 \text{ joules}}{0.653 \text{ ml}} = 3.68 \times 10^3 \text{ joules/ml.}$$ ◀

▶ *Example 2*

Instead of making an experiment, as above, one may prefer to calculate the conversion factor from highly accurate available data. Given 0.91671 gm/ml and 0.99987 gm/ml as the densities at 0°C of ice and water respectively, and 79.72 cal/gm as the heat of fusion of ice,[7] calculate the conversion factor:

$$\alpha' = \frac{\text{Number of calories of heat}}{\text{Number of ml volume change}}.$$

Solution. What will be the volume change when 1 gm of ice is melted? This we can most easily find by converting the given densities to the corresponding specific volumes, i.e., the volumes occupied by one gram of each material.

$$\begin{aligned}
\text{Spec. vol. ice} &= 1/\text{density ice} = 1/0.91671 \text{ gm/ml} = 1.09086 \text{ ml/gm}\\
\text{Spec. vol. water} &= 1/\text{dens. water} = 1/0.99987 \text{ gm/ml} = 1.00013 \text{ ml/gm}\\
\therefore \text{Volume change per gm of ice melted} &= 0.09073 \text{ ml/gm}
\end{aligned}$$

But this is the volume change that results whenever the ice calorimeter receives the 79.72 cal required to melt one gm of ice. Therefore,

$$\alpha' = \frac{79.72 \text{ cal/gm}}{0.09073 \text{ ml/gm}} = 878.7 \text{ cal/ml.}$$

Is the figure obtained in the first example consistent with this more accurate result? ◀

▶ *Example 3*

Just 0.01 mole of a compound was dissolved at 0°C in 10 ml of water in chamber R of an ice calorimeter. The consequent change in the volume reading shown on tube C was an *increase* of 0.0284 ml. Determine the direction and magnitude of the heat transfer when 1 mole of the compound is dissolved at 0°C in 1 liter of water.

Solution. Using the factor calculated in the last example, we see that the heat transfer in the actual calorimetric experiment was

$$0.0284 \text{ ml} \times 878.7 \text{ cal/ml} = 25.0 \text{ cal}.$$

Observe that the volume *increased:* this means that in the calorimeter ice was formed at the expense of the less voluminous water. Thus the direction of heat transfer was from the calorimeter to the solution in R. If the solution represents our "system," the heat transferred in this endothermic process will carry a positive sign. Hence the heat of solution of 1 mole of compound in 1 liter of water will be

$$+25.0 \times 100 = +2.50 \times 10^3 \text{ cal/mole}.$$ ◀

Heat is thus seen to be readily measurable, and the indicated style of measurement is itself a definition of what we mean by "heat." Turning then to "work," we can be much briefer. The only species of work of major concern to us is the familiar mechanical work of which we write:

$$\underset{\text{(joules)}}{\text{Work}} = \underset{\text{(newtons)}}{\text{Force}} \times \underset{\text{(meters)}}{\text{Distance}}.$$

Toward the end of this text we touch briefly upon electrical work, for which we write:

$$\underset{\text{(joules)}}{\text{Work}} = \underset{\text{(volts)}}{\text{Electromotive force}} \times \underset{\text{(coulombs)}}{\text{Charge transported}}.$$

Now both kinds of work can be used to lift weights—with 100% efficiency in principle, i.e., using frictionless pulleys, ideal electric motors, etc. We can measure (and to that extent define) "work" in terms of this effect.

We shall say that work has been done *by* a system, on its surroundings, if weights are observed to rise in the surroundings as a result of interaction with the system. And we shall say that work has been done *on* a system, by its surroundings, if weights are observed to descend in the surroundings as a result of interaction with the system. From the observed rise or fall, the work transfer can at once be calculated. Suppose that, in a field where the gravitational acceleration is g, a particular mass M is observed initially at height h_i and finally at height h_f. Then:

$$\underset{\text{(joules)}}{\text{Work}} = \text{Vertical force} \times \text{Vertical distance}$$
$$= \text{Mass} \times \text{Gravitational acceleration} \times (h_f - h_i)$$
$$= M \text{ (kgm)} \times g \text{ (meters/sec}^2) \times \Delta h \text{ (meters)}.$$

Observe that in $\Delta h (\equiv h_f - h_i)$ the operator Δ is used as it always will be—demanding that we form the difference [final value of indicated parameter *minus* the initial value thereof]. The sign convention for work given on p. 4 is then taken care of automatically. For when the weight rises Δh will be *positive* and, indeed, to make the weight rise, work must have been done *by* the system. And when the weight descends Δh will be *negative* and, indeed, through the descent of the weight, work will have been done *on* the system. Thus a value for work (w) correct in both sign and magnitude is given by the simple relation:

$$w = Mg\,\Delta h. \tag{a}$$

Ample illustration of work calculations will be found in the next section, which introduces one of the key concepts of thermodynamics.

REVERSIBILITY

The amount of work extractable from even the simplest process of change depends markedly on just how that process is conducted. This point we highlight by subjecting the set-up sketched in Fig. 3 to a style of analysis first suggested by Eberhardt.[8]

Let our *system* be a long spring, suspended from a hook at its upper end and bearing a light pan at its lower end. Let the *surroundings* be all the rest of the world, and most particularly a group of weights totaling 1600 gm. As state I of the system we have the spring stretched to the length at which it supports 1600 gm in the pan, as shown. As state II of the system we have the spring relaxed to the length at which it supports the empty pan. Assuming the spring ideal in its conformity to Hooke's law—and many real springs closely approach the ideal—the elongation of the spring will be directly proportional to the load it sustains. In that case, if the elongation produced by 1600 gm is the 81.6 cm shown at the extreme left, the elongation produced by 800 gm will be 40.8 cm; that produced by 400 gm will be 20.4 cm; and so on. We then construct a set of nine shelves at the heights shown at the extreme right, measured from the level of the pan when, with a 1600-gm load, the system stands in state I. And now the question: how much work can be obtained from this system when it passes from state I to state II?

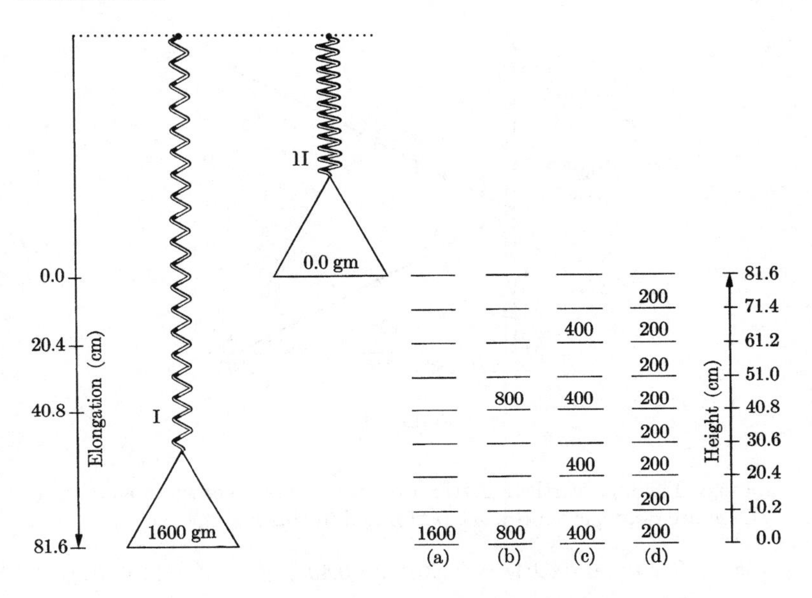

FIGURE 3

Possibility (*a*). Temporarily securing the pan, we slide all 1600 gm horizontally onto the bottom shelf. On then releasing the pan, we see the system pass through a series of progressively less violent oscillations until at last it settles down in state II. Now all the weights remain at precisely their original level, so that no work whatever has been accomplished. Thus we write $\vec{w}_a = 0$, where the arrow signifies the forward change from state I to state II, and the subscript indicates the first of a number of possible ways of conducting that change.

Possibility (*b*). After securing the pan with its 1600 gm, we transfer to the bottom shelf only 800 gm. When released, the pan now comes to rest at the level of the middle shelf, at height 40.8 cm. We then slide the remaining 800 gm horizontally onto this shelf, after which the system passes on into state II. In this case 800 gm (0.80 kgm) has been lifted through 40.8 cm (0.408 meter). Taking 9.8 meter/sec^2 as the normal gravitational acceleration, equation (a) yields

$$\vec{w}_b = (0.80)(9.8)(0.408) = 3.2 \text{ joules.}$$

Possibility (*c*). This time we remove the weights in 400-gm lots, transferring each lot horizontally onto the shelf level with which the pan is then

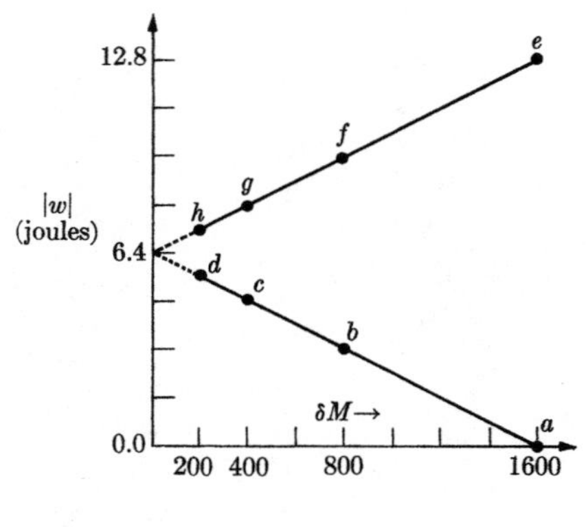

FIGURE 4

standing. The final vertical distribution of weights is shown in column (c) of the figure, and the total work performed in this instance is

$$\vec{w}_c = (0.40)(9.8)(0.204) + (0.40)(9.8)(0.408) + (0.40)(9.8)(0.612)$$
$$= 3(0.40)(9.8)(0.408) = 4.8 \text{ joules.}$$

Possibility (*d*). Removing the weights in 200-gm lots, we find the final distribution shown in column (d), and the total work done is

$$\vec{w}_d = (0.20)(9.8)(0.102) + \cdots = 7(0.20)(9.8)(0.408) = 5.6 \text{ joules.}$$

Plotting for each trial the absolute value of the work done *versus* the mass decrement (δM) used in that trial, we obtain the suggestive lower line in Fig. 4.

Observe that: (i) the smaller the mass decrement used, the greater the work recovered; and (ii) in the limit $\delta M \rightarrow 0$ one would garner the maximum work *output* (→6.4 joules) recoverable from the passage of the system, along the indicated path,* from state I to state II. These results are not without interest, but their full significance emerges only when we compare them with the work *input* required to return the system from state II to state I. In Fig. 5 we display a few of many ways in which the 1600 gm can be arranged to bring about the restoration of state I.

Possibility (*e*). Here we envision the system in state II and a total of 1600 gm placed on the topmost of the auxiliary shelves. We move the entire

* As noted on p. 22, the limiting work along a *different path* between states I and II might be quite different.

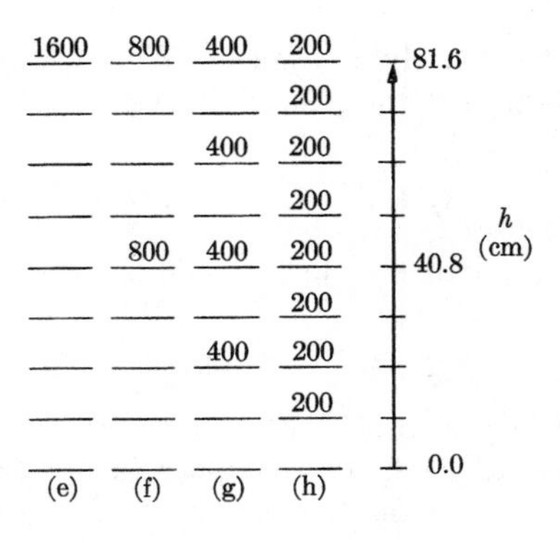

FIGURE 5

1600 gm horizontally onto the pan, which is temporarily locked in place. Releasing it, we see the system pass through a series of progressively less violent oscillations until at last it settles down in state I. In this restoration of state I, 1600 gm has *descended* 81.6 cm, i.e., $\Delta h = -81.6$ cm. Symbolizing by $\overleftarrow{w}$ the work *input* required to bring about this inverse change from II to I, we write

$$\overleftarrow{w}_e = (1.60)(9.8)(-0.816) = -12.8 \text{ joules.}$$

This is twice the magnitude of the maximum work *output* recoverable when the system passes from state I to state II. Is there not some more economical way of restoring state I?

Possibility (f). This time we make a beginning with 800 gm on the top shelf and 800 gm on the middle shelf. Starting with the system locked in state II, we slide the 800 gm from the top shelf horizontally onto the balance pan. When the system is released, the pan comes to rest at the level of the middle shelf. Then transferring the 800 gm from that shelf to the pan, we bring the system back into state I. In the entire process, 800 gm descends 81.6 cm, and a further 800 gm descends 40.8 cm. The total work input is, in this case,

$$\begin{aligned}\overleftarrow{w}_f &= (0.80)(9.8)(-0.816) + (0.80)(9.8)(-0.408)\\ &= 3(0.80)(9.8)(-0.408) = -9.6 \text{ joules.}\end{aligned}$$

This is a more economical way of restoring state I, and the way to still greater economy is at once evident.

Possibility (g). We begin here with 400-gm masses distributed on the shelves as shown in panel (g) of the last figure. These masses are successively transferred horizontally to the pan. When the system has thus been restored to state I, all these masses will have descended to the level of the bottom shelf. Hence:

$$\overleftarrow{w}_g = (0.40)(9.8)(-0.816) + \cdots = 5(0.40)(9.8)(-0.408) = -8.0 \text{ joules.}$$

Possibility (*h*). An analysis which should by now be obvious yields for this last case

$$\overleftarrow{w}_h = 9(0.20)(9.8)(-0.408) = -7.2 \text{ joules.}$$

Plotting for each trial the absolute value of the work input, *versus* the mass increment in that trial, we obtain the upper line in Fig. 4.

Observe that: (i) the smaller the mass increment used, the less the work input required to restore state I; and (ii) in the limit $\delta M \to 0$ one attains the minimum work input ($\to -6.4$ joules) required to return the system, along the indicated path, from state II to state I. But the all-important conclusion of the entire analysis is this: *In the limit $\delta M \to 0$, the minimum work input required to bring the system from state II to state I is precisely compensated by the maximum work output recovered when the system changes, along the same path, from state I to state II.* Processes that meet this limiting condition we term "reversible," and we use the subscript rev to distinguish the corresponding limiting value of a work term. We can then give very brief algebraic expression to the italicized statement just above:

$$\overrightarrow{w}_{\text{rev}} + \overleftarrow{w}_{\text{rev}} = 0. \tag{b}$$

Or, again in words, actual reversal of the direction of a reversible change alters only the sign, *not* the magnitude of the work term.

As elsewhere in our exploration of the vocabulary of thermodynamics, through analysis of a single example we seek here to highlight the most important aspects of a *general* concept: reversibility. The first and most obvious sense of the word arises from the perfect matching of work terms we have expressed in equation (b). When a process has been conducted reversibly we can, by also performing the inverse process reversibly, set the system back in precisely its initial state, with *no* net expenditure of work in the overall bipartite process. What we have done is "reversible" just because, at the end, the world (i.e., the system *and* its surroundings) is once again as it was in the beginning.

Now the maximization of work output and minimization of work input required for reversibility is attainable only in the special case that the tension of the spring and the load on the pan are matched to within an infinitesimal margin. In these circumstances, an infinitesimal alteration in the surroundings must suffice not merely to arrest but to *reverse* the direction of the change. Thus when the contraction of the spring proceeds reversibly, that contraction will be reversed if at any point we *add* (rather than withdraw) an infinitesimal mass $\delta M \rightarrow 0$. This proves to be a highly significant feature of reversible change, for it turns out that there are *some* changes, like the frictional conversion of work into heat, that can in *no* circumstances be reversed by a small alteration in the surroundings.

Reversibility thus presumes the absence of frictional forces. But, even in an ideal frictionless system, a force only infinitesimally superior to load can yield no more than an infinitesimal rate of change. Hence another distinctive characteristic of reversible change is that it must proceed immeasurably slowly. In effect, a reversible change proceeds by way of an infinitely prolonged succession of pseudo-equilibria, each no more than infinitesimally distant from a true equilibrium state. Though (as noted on p. 2) in systems undergoing rapid change, unique values *cannot* be assigned to temperature and pressure, these parameters remain as well defined during reversible change as they are at equilibrium. Precisely this resolution of change, into a series of well defined quasi-static states, is what brings change within the scope of a classical thermodynamics which —exclusively concerned as it is with equilibria and not with rates—is perhaps better denominated thermostatics.

For reversibility the basic requirement (typified by $\delta M \rightarrow 0$) thus entails further conditions so exacting that we can be quite certain of one thing. *No observable change*, proceeding by definition at a finite rate, *is strictly reversible*. The perfectly reversible change is the unobservable ideal limit approached, but never quite attained, by some observable changes. Yet often we can find out exactly what would happen in a reversible change: see how we have fully established the work term that *would* characterize the reversible stretching or contraction of the spring in our example. In this the concept of reversibility is as concrete though abstract, as powerful though simple, as the mathematical concept of a limit!

PRESSURE-VOLUME WORK

In expanding, any system does work by pushing back the atmosphere around it; in contracting, any system is the recipient of work done by the atmosphere pressing upon it. Lest the atmosphere seem too impalpable to play the role of receiver or donor of work, we replace it with an easily visualizable equivalent. But all the results now to be derived apply *generally*—to any system that changes its volume in direct contact with the atmosphere.

As system, consider a cylinder charged with a fluid (e.g., an ideal gas) and fitted with a hypothetically frictionless piston of negligible mass. Pumping off all the surrounding air, which ordinarily presses on the piston, we produce the same effect by setting a suitable mass on the shelf shown atop the piston rod in Fig. 6. If the normal external pressure is symbolized by P_x, and the cross-sectional area of the piston by A, the mass (M) to

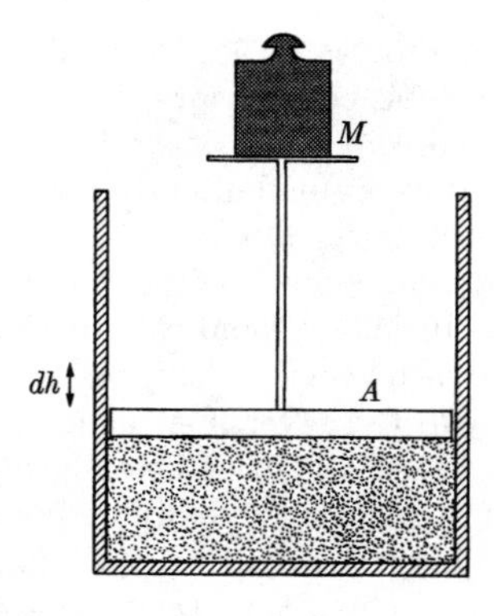

FIGURE 6

be used must satisfy the equation:

$$\frac{Mg}{A} = \frac{\text{Force}}{\text{Area}} \equiv \text{Pressure} \equiv P_x.$$

How much work is done by or on the system when the piston moves some minute vertical distance dh? Starting from equation (a), we write

$$\text{Work} = Mg\,dh = \frac{Mg}{A}\,A\,dh = P_x\,dV,$$

where dV symbolizes the infinitesimal volume increment clearly represented by the product $A\,dh$. Observe that, in accordance with our convention, the work term will be positive when the system expands (dV positive) and negative when the system contracts (dV negative).

How shall we calculate the work term when the piston moves a finite distance? Even if P_x changes, we have only to break the change into a series of infinitesimal steps in each of which P_x can be treated as effectively constant. Summing over all the steps that bridge between the initial volume (V_1) and the final volume (V_2), we find:

$$w = \sum P_x\,dV = \int_{V_1}^{V_2} P_x\,dV.$$

This can be drastically simplified if P_x remains *constant* throughout the change:

$$w = \int_{V_1}^{V_2} P_x\,dV \underset{\text{IF } P_x \text{ is constant}}{=} P_x \int_{V_1}^{V_2} dV = P_x(V_2 - V_1) = P_x\,\Delta V, \qquad (1)$$

where ΔV symbolizes a finite change of volume. Should not the pressure of the gas inside the system somehow enter into the expression for work? Intuitively we feel that when the initial inside pressure is large, more work should be recoverable from the system than when the initial inside pressure is small. And indeed this is so: the high-pressure gas yields a greater expansion (ΔV) against the fixed external pressure. Thus the pressure of the gas inside the system *is* duly reflected in the magnitude of the work term. But, with the system defined as we have defined it, only the *external* pressure appears as such in our expression for pressure-volume work.[9]

If the external pressure is expressed in atmospheres, and the volume change of the system in liters, the calculated work will be expressed in the unsystematic but very useful unit of liter-atmospheres (abbreviated, lit-atm).

▶ *Example 4*

How many joules are equivalent to one lit-atm of work?

Solution. For the 76.00-cm column of mercury (density 13.596 gm/cm^3) supported by the atmosphere,

$$M/A = (76 \text{ cm})(13.6 \text{ gm/cm}^3) = 1033 \text{ gm/cm}^2 = 1.033 \text{ kgm/cm}^2.$$

Hence

$$P_x = Mg/A = (1.033 \text{ kgm/cm}^2)(9.81 \text{ meter/sec}^2)$$
$$= 10.13 \text{ kgm-meter/sec}^2\text{-cm}^2 = 10.13 \text{ newton/cm}^2;$$

and one lit-atm will represent

$$P_x \, \Delta V = (10.13 \text{ newton/cm}^2)(1000 \text{ cm}^3) = 10130 \text{ newton-cm}$$
$$= 101.3 \text{ newton-meter} = 101.3 \text{ joule.}$$ ◀

Isothermal expansion of an ideal gas

To extend our view of pressure-volume work, while also consolidating our grasp of "reversibility," let us consider in some detail the isothermal expansion of an ideal gas described by the equation $PV = nRT$. Constant temperature is the sense of "isothermal" (Gr. *isos*, same + *thermē*, warmth). In an "isothermal process" the system must be so placed (e.g., in a thermostat) that heat transfers will occur as necessary to hold invariant the temperature of the system. Suppose we thus "thermostat" a system made by charging our cylinder-piston device with enough ideal gas to exert 10 atm pressure in a 1-liter volume. How much work will be done when the gas expands until it occupies a 10-liter volume at a pressure of 1 atm?

Consider first the case in which *no* mass rests on the shelf of the piston assembly which, recall, is itself of negligible mass. The piston is then held against the 10-atm interior pressure only by a lower catch that forms part of the system. When this catch is released, the piston is driven back against zero external pressure until again arrested—by an upper catch, set to stop

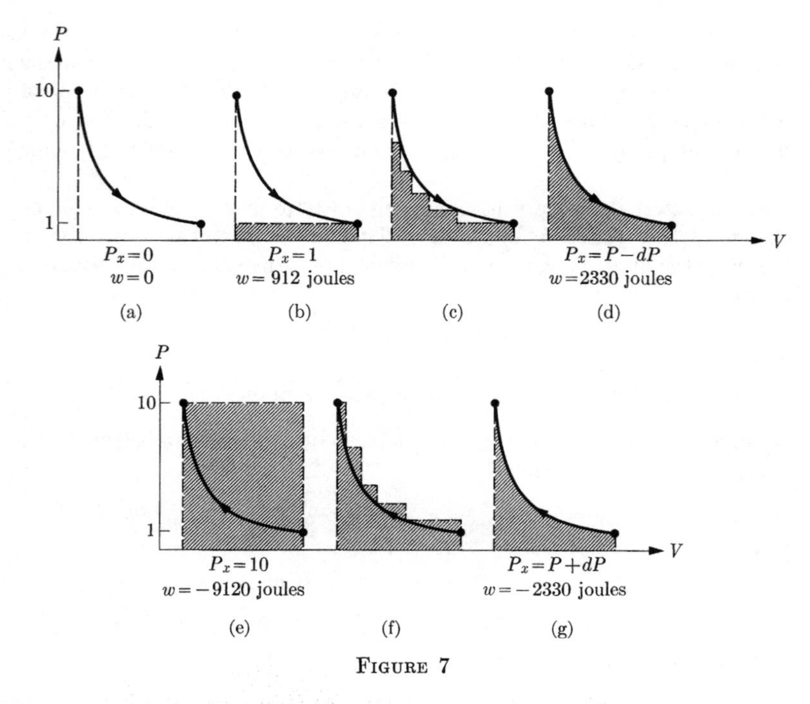

FIGURE 7

the expansion when the internal volume reaches 10 liters. We have no knowledge of the interior temperature and pressure of the gas *during* this highly irreversible process. Only when the expansion has *ended*, and the gas has come to equilibrium, do we know that its state must be represented by the second point marked on the Boyle's-law curve shown in Fig. 7(a). In this case the line $P_x = 0$ coincides with the horizontal axis, and *no* work is performed because

$$\vec{w}_a = P_x \, \Delta V = (0)(9) = 0.$$

A second possibility would be to place on the piston shelf a mass just sufficient to act upon the piston like an external pressure $P_x = 1$ atm. When the piston is released from the lower catch, the expansion thus initiated ends only when the gas occupies a 10-liter volume at a pressure of 1 atm. Once again the Boyle's-law curve shown in Fig. 7(b) does *not* represent the state of the gas during the irreversible expansion. But when temperature- and pressure-equilibrium have been reestablished at the end of the expansion, the state of the gas inside the system is that defined by the intersection of the Boyle's-law curve with the line $P_x = 1$ atm. Indeed the expansion ends precisely because the internal pressure here falls to equality with the external pressure. The work done in the expansion is indicated in panel (b) by the shaded area, which clearly represents $P_x\,\Delta V$. Algebraically,

$$\vec{w}_b = P_x\,\Delta V = (1)(9) = 9 \text{ lit-atm} = 912 \text{ joules.}$$

An even better recovery of work is clearly possible if we are willing to readjust P_x from time to time, as shown in panel (c). We then begin with a piston load that simulates an external pressure of, say, 8 atm. Once released, the piston recedes until at last it comes to rest when the internal volume corresponds to the intersection of the Boyle's-law curve with the line $P_x = 8$ atm. The internal and external pressures are here equal, and to set the piston in motion again we must reduce the mass acting on it to, say, the equivalent of $P_x = 6$ atm. There follows a further expansion, again terminated when the pressure inside the system has fallen to equality with P_x. We then make a further reduction in P_x, . . . and so on, so that P_x describes a descending zig-zag bounded by the Boyle's-law curve as shown in panel (c). The shaded area represents the total work recovered, which is easily determined by actually summing the $P_x\,\Delta V$ products for all the rectangular segments. The work recovered is obviously much larger than 912 joules; yet, equally obviously, still less than the *maximum* work recoverable from the isothermal expansion. But of this maximum work we recover an ever greater proportion by increasing the number of steps through which we reduce P_x.

The greater the number of steps, the more closely does the step curve approximate the smooth curve, and the more closely does $\sum P_x\,\Delta V$ for all the rectangular segments approach the area under the smooth curve. In the limit, when P_x is reduced through an infinite series of infinitesimal steps, the step curve becomes indistinguishable from the smooth curve—and one would recover the maximum work, represented by the shaded area in panel (d). This area we will calculate presently, by a simple application of the calculus developed in Appendix II. But let us first consider the compression that is the inverse of the expansion just discussed.

To recompress the enclosed gas, from 10 liters at 1 atm to 1 liter at 10 atm, it is simplest to place atop the piston a mass equivalent to $P_x = 10$ atm. The piston is then sustained in place only by the upper catch, on release of which it falls with some violence. Ultimately the compression ceases when—at the intersection of the Boyle's-law curve with the line $P_x = 10$ atm—the internal pressure rises to equality with P_x as the volume of the enclosed gas is reduced to 1 liter. The work expended in this compression is indicated by the shaded area in Fig. 7(e), which represents the product $P_x\,\Delta V$. Algebraically,

$$\overleftarrow{w}_e = P_x\,\Delta V = (10)(-9) = -90 \text{ lit-atm} = -9120 \text{ joules.}$$

The same compression is evidently attainable with a much smaller expenditure of work, if we are prepared to readjust P_x from time to time. When we start with P_x only slightly greater than 1 atm, the compression is soon arrested at the intersection of the Boyle's-law curve with the P_x line. To induce a further compression, we must somewhat increase P_x by adding to the mass acting on the piston. At the close of the next compressive step, P_x must again be increased, . . . and so on as shown in panel (f) where, again, the shaded area indicates the work expended.

Still further reduction in the expenditure of work is clearly feasible. The greater the number of steps through which we increase P_x, the more closely does the ascending step-curve approximate the smooth curve. In the limit where P_x is increased through an infinite series of infinitesimal steps, the step-curve becomes indistinguishable from the smooth curve, and one expends the minimum work represented by the shaded area in panel (g).

Now surely the limiting cases depicted in panels (d) and (g) are reversible processes that meet all the conditions listed in our earlier analysis of reversibility. First and foremost, we have found that, for isothermal expansion and compression respectively, maximum work output and minimum work input are so represented by the same area that, in line with equation (b), we have here:

$$\overrightarrow{w}_d + \overleftarrow{w}_g = 0.$$

Second, with at most an infinitesimal margin (dP) between external and internal pressure, the *direction* of change can easily be reversed by an infinitesimal change in the surroundings, i.e., a transfer of mass δM to or from the top of the piston. Third, driven only by the infinitesimal unbalance represented by dP, the reversible change proceeds so slowly that the state of the enclosed gas never deviates significantly from the continuous series of strictly isothermal equilibrium states defined by the Boyle's-law curve. As noted earlier, this reduction of change to a series of pseudo-equilibrium states plays a vital role in thermodynamic analysis. This role we now illustrate while at the same time completing our examination of the isothermal expansion and compression of the ideal gas.

We have still to determine the magnitude of the reversible work (w_{rev}). For a finite change of volume, from V_1 to V_2, our general expression for the pressure-volume work reads:

$$w = \int_{V_1}^{V_2} P_x \, dV.$$

Actually to perform the integration, we must now discover some way to express P_x as a function of V.

In a reversible change, P_x never differs more than infinitesimally from the interior gas pressure (P). For such a change,

$$P_x = P \pm dP,$$

where the minus and plus signs refer respectively to expansions and compressions. (Why?) Substituting in our general expression for w, we obtain for the special case of a reversible change

$$w_{\text{rev}} = \int_{V_1}^{V_2} (P \pm dP)\, dV = \int_{V_1}^{V_2} (P\, dV \pm dP\, dV).$$

The term $dP\, dV$ is an infinitesimal of higher order: no error attends its complete suppression, leaving

$$w_{\text{rev}} = \int_{V_1}^{V_2} P\, dV.$$

Now we have only to express the *internal* pressure as a function of V. Recalling that $PV = nRT$ is the equation of state* for the ideal gas within the system, we write

$$w_{\text{rev}} = \int_{V_1}^{V_2} \frac{nRT}{V}\, dV.$$

In the *isothermal* change of a *fixed* quantity of gas, the term nRT is a constant we can move across the integral sign. Thus we arrive at a familiar simple integral:

$$w_{\text{rev}} = nRT \int_{V_1}^{V_2} \frac{dV}{V} = nRT \ln \frac{V_2}{V_1}.$$

To this we may add two equivalent expressions that follow from Boyle's law and the relation of natural to denary logarithms:

$$w_{\text{rev}} = nRT \ln \frac{V_2}{V_1} = 2.30\, nRT \log \frac{V_2}{V_1} = 2.30\, nRT \log \frac{P_1}{P_2}. \qquad \text{(c)}$$

These expressions apply to every reversible isothermal expansion or compression of a fixed quantity of ideal gas. In the concrete example discussed above, we never specified the temperature of the isothermal change, but we know that the gas occupies 1 liter at 10 atm pressure. Therefore $nRT = PV = (10)(1) = 10$ lit-atm and, for the reversible expansion from 1 to 10 liters, we write

$$\vec{w}_{\text{rev}} = 2.30\, nRT \log \frac{V_2}{V_1} = 2.30(10) \log \frac{10}{1} = 23 \text{ lit-atm} = 2330 \text{ joules}.$$

---

* The appropriate value for R will of course vary with the units used to express V and P. If liters and atmospheres are the units, then

$$R = \frac{(1 \text{ atm})(22.4 \text{ lit})}{(1 \text{ mole})(273°\text{K})} = 0.082 \text{ lit-atm/mole-°K}.$$

The dimensions of R are those of energy/mole-°K, and in other energy units R is ~8.3 joules/mole-°K, ~2.0 cal/mole-°K, etc.

For the corresponding compression,

$$\overleftarrow{w}_{\text{rev}} = 2.30(10) \log \tfrac{1}{10} = -2.30(10) \log 10 = -23 \text{ lit-atm} = -2330 \text{ joules}.$$

Note how the appropriate signs are automatically associated with the work terms if we take care to substitute for V_1 and V_2 the initial and final volumes respectively. In an expansion $V_2 > V_1$, and both $\log V_2/V_1$ and w are positive; in a compression $V_2 < V_1$, and both $\log V_2/V_1$ and w are then negative.

States, paths, and constraints

In our analysis of the isothermal expansion of the ideal gas, as in our analysis of the Hooke's-law spring, we found $\overrightarrow{w} \leq \overrightarrow{w}_{\text{rev}}$. Here $\overrightarrow{w}$ symbolizes the work output associated with the forward change in the general case, and $\overrightarrow{w}_{\text{rev}}$ symbolizes the work output in the special case of a *reversible* progress of that same change along the same path. Because of the negative sign attaching to all the corresponding $\overleftarrow{w}$ terms, for them too we can write $\overleftarrow{w} \leq \overleftarrow{w}_{\text{rev}}$. Regardless of whether a work input or a work output is involved, for a given change conducted along a given path, we find that

$$w \leq w_{\text{rev}}, \tag{d}$$

where the equality applies in the special limiting case of reversibility, and the inequality applies in all other cases.

Why must we insist that w and w_{rev} refer to the *same change* proceeding along the *same path*? Why they must refer to the same change is plain enough: there need be no relation whatever between w measured for one change in one system, and w_{rev} measured for an entirely different change and perhaps also on an entirely different system. Why restriction to the

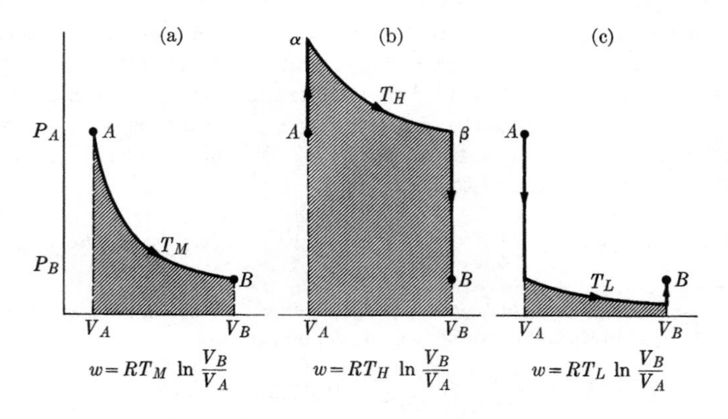

FIGURE 8

same path is necessary may be less evident. We explore that question now, in the specific context of the expansion of an ideal gas.

Let our system be the familiar cylinder-piston, containing now just one mole of ideal gas. A single point on a plot of P *vs* V suffices fully to specify the state of the gas: given that $PV = nRT$, a unique value of T is entailed as soon as the values of P, V, and n are fixed. Consider now any two points (A and B) for which the temperature (T_M) is the same. To say that we are concerned with the change from A to B is then amply to specify the change at issue but, as Fig. 8 suggests, wholly insufficient to establish along which of innumerable possible paths that change is to proceed. Under the line representing each path of change, the shaded area represents the maximum work recoverable when the expansion from A to B follows that path: note the gross variation of this maximum from one path to another.

Path (a) is the simple isothermal expansion discussed in the last section, where we found that the work associated with this path is

$$w_a = RT_M \ln V_B/V_A.$$

Panel (b) displays a more complex three-step path. Starting with the gas in state A, and the piston locked in place, we warm the system from T_M to some higher temperature T_H. With the volume held constant at V_A, in this first step the pressure rises to the value indicated by point α. In the second step, a reversible isothermal expansion, at temperature T_H, carries the system from α to point β, where the volume is V_B. In the third step, cooling at constant volume, from T_H to T_M, at last brings the gas to the state represented by point B. Now the initial and final constant-volume segments contribute nothing to the production of work along this three-step path, because for them $w = \int P_x\, dV = 0$. Hence the total work (w_b) recoverable along this path is just that produced in the isothermal expansion, which means

$$w_b = RT_H \ln V_B/V_A.$$

The situation depicted in panel (c) is basically the same as that in (b). But where in (b) we first heated the system to some higher temperature T_H, and at the end cooled it back to T_M, in (c) we first cool the system to some lower temperature T_L, and at the end warm it back to T_M. Once again the total work (w_c) reduces to that yielded by the isothermal expansion, so that

$$w_c = RT_L \ln V_B/V_A.$$

Now to precisely the extent that $T_H > T_M > T_L$, so also $w_b > w_a > w_c$. Yet w_a, w_b, and w_c are all quite properly denominated w_{rev}. Thus we see that w_{rev} is a well-defined quantity only for a specified change proceeding along a specified path. That is, when we stipulate that a change is to proceed "reversibly," we do not thereby specify a path but only the *manner* in which any path is to be traversed, e.g., with force and load matched to within an infinitesimal margin, etc. When, on the other hand, we say that the entire change is to proceed "isothermally," we do thereby specify a path, and w_{rev} for the isothermal change is a well-defined quantity.[10] For when we stipulate the *isothermal* expansion of an ideal gas, we invoke a constraint that excludes all possible paths save one, (a), with which is associated a unique value for w_{rev}.*

Like the specification of constant temperature, certain other constraints may operate to exclude all but one path for the given change. Constant-volume and constant-pressure requirements are such constraints. When only pressure-volume work is possible, we have seen that a constant-volume process must proceed with $w = \int P_x\,dV = 0$; and a constant-pressure process will involve $w = \int P_x\,dV = P_x \int dV = P_x\,\Delta V$. The usefulness of such processes may well seem impossibly narrow. After all, only initial and final states that are alike in volume can be linked by a constant-volume path. And a constant-pressure path can be used to link only initial and final states that are alike in pressure. But *any* initial and final states, represented by points in a P–V plane, can be joined by two-step paths made up of coordinated constant-volume and constant-pressure segments. Such a possibility invests us with significant conceptual powers—even though, as indicated in panels (a) and (b) of Fig. 9, variation in the *order* of the steps yields two different paths with which are associated different values for w_{rev}.

To the three constraints so far noted, we may add one more that complements the isothermal constraint in somewhat the manner that constant-volume and constant-pressure constraints complement each other. Where an isothermal process presumes whatever heat transfers are required to hold the temperature constant, in a process that is "adiabatic" (Gr. *a*, not + *dia*, through + *bainein*, to go) no heat whatever is permitted to enter or leave the system. If we stipulate that the reversible expansion of an ideal gas shall be adiabatic, we specify a path as unique as when we stipulate that the expansion shall be isothermal. On p. 54 we show that along the adiabatic path the pressure falls off rather more steeply, with increasing volume, than in the corresponding isothermal expansion. By virtue of this difference, we can join *any* two points in the P–V plane

---

* Like the pressure of a gas, the stiffness of a spring is altered by variation of its temperature; and w_{rev} for the *non*isothermal contraction of the spring is rendered ambiguous by the same kinds of alternatives displayed for a gas in Fig. 8. A tacit assumption of isothermal change is thus what underlies the qualification "by the indicated path," which appears repeatedly in our earlier analysis of reversibility, beginning on p. 10.

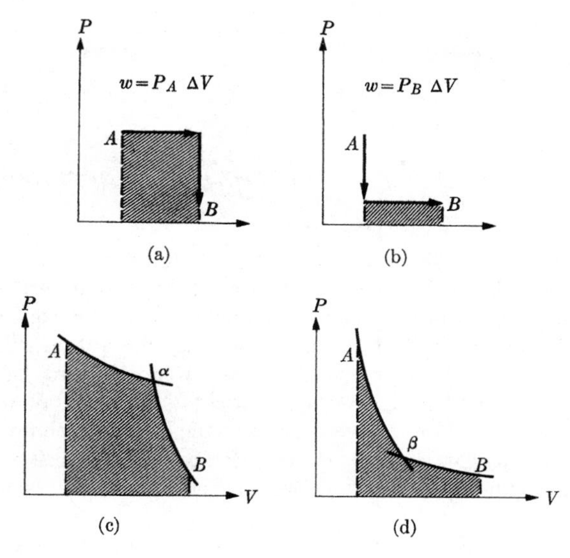

FIGURE 9

with two-step paths made up of coordinated adiabatic and isothermal segments. In panel (c) of Fig. 9 the gas first expands isothermally from state A to state α, the latter being so chosen that a further adiabatic expansion carries the gas to state B. In panel (d), on the other hand, there is first the adiabatic expansion from A to β, where β is so chosen that a further isothermal expansion brings the gas to state B. As before, variation in the *order* of the two steps yields two different paths, with which are associated what the shaded areas show to be very different values for w_{rev}. And so we come back to where we began. When a system undergoes a given change of state, the maximum amount of recoverable work varies markedly with the path along which the change proceeds. And, as the equation $w \leq w_{rev}$ implies, the amount of work actually recovered (w) is even less well defined: like w_{rev} dependent on the identity of the change and the path thereof, w depends in addition on the manner in which the path of change is traversed.

2

The First Principle of Thermodynamics

Heat and work are fundamental concepts, but often the least convenient of concepts with which to conduct thermodynamic analyses. A nuisance in itself, the dependence on the path of change shown by work transfers—and, as we will see, by heat transfers also—signifies a deeper disability in the concepts of heat and work. Heat and work are birds of passage—never found as such in residence. Quantities of heat and work are perfectly determinate *as* they are transferred across the boundary that delimits the system. But *after* completion of the transfers we cannot speak of the system (or of its surroundings) as having some new heat content or some new work content.

Consider work. Suppose that some definite work content were associated with each state of a system. In that case the change of the system, from any given initial state to any given final state, must be accompanied by a definite, invariant transfer of work to or from the system. But in the last chapter we found that, during the passage between any two given states of a system, the work transferred depends markedly on the path of change —not to mention also the manner in which that path is traversed. Thus observation shows variability where the hypothesis of work content demands invariance. We must then relinquish the hypothesis, and so we conclude that a definite work content *cannot* be associated with each state of a system.

Consider heat. As a system let us take some quantity of water, say, at a temperature of 0°C. By inserting in the water a resistance coil carrying an electric current, we transfer to the system enough *heat* to raise its temperature to, say, 1°C. But, as J. P. Joule was the first to show convincingly, we can bring the system to exactly the same final state by a transfer of *work* made, perhaps, by inserting in the water a paddle wheel driven by the fall of weights as shown in Fig. 10. Here then, as in numberless other instances, a system undergoes precisely the same change of state when it traverses one path that involves a transfer of *heat only,* and when it traverses instead another path that involves a transfer of *work only*. But this is truly dramatic evidence that for heat, as for work, the amount transferred in a given change of state varies with the path of change. And then the same form of argument used in the last paragraph

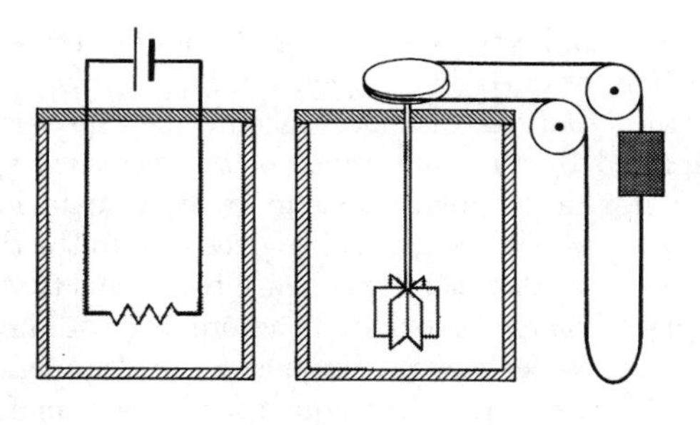

FIGURE 10

will yield here a conclusion of the same form: a definite heat content *cannot* be associated with each state of a system.

We simply cannot say of a system that in a certain state it contains a certain quantity of work, or a certain quantity of heat. Thus heat and work are not "functions of state" of the sort that have proved most powerful as tools of thermodynamic analysis. Such a function must (i) have a definite value for a given system in a given state, independent of the history of that system; and (ii) change by a definite amount in a given change of state, independent of the path along which that change proceeds. Of five functions that meet these two overlapping criteria, the first we shall discuss is *energy*—the conservation, or constancy, of which is asserted by the first principle of thermodynamics.[11]

ENERGY

Consider a volume of water stationary in a pool at the head of a waterfall. It has what we may call "privilege of position," in that once it has dropped over the fall we must do work to return it to its original position. As the water passes over the fall its "privilege of position" vanishes, but at the same time it acquires *vis viva*, the "living force" of motion. By passing the water through a turbodynamo, we strip it of its *vis viva* and simultaneously acquire electric power which, vanishing when the dynamo is shorted through a resistance, there gives rise to an evolution of heat. If the water drops directly to the bottom of the fall, without passing through the turbine, *vis viva* disappears without the production of electric power; but at the bottom of the fall the water has a temperature slightly higher than that with which it left the top of the fall—just as though it had received the heat from the above-noted resistor. Now *a priori* there is *no* reason to suppose that "privilege of position," *vis viva*, electric power, and heat—qualitatively apparently utterly different—stand in any relation whatever to each other. Experience, however, teaches us to regard them all as diverse manifestations of a single fundamental potency: energy (Gr. *energos*, active; from *en*, in + *ergon*, work).

We have learned that the disappearance of one "species of energy" is always accompanied by the appearance of another such species—as, in our example, mechanical potential energy gives way to kinetic energy, which gives way to electric energy, which gives way to the thermal energy we call heat. Far more than this, experience teaches us that the interconversions of species of energy take place in accordance with invariant ratios. That is, when mechanical energy disappears, an "equivalent" quantity of electric energy, or an "equivalent" quantity of heat, makes its appearance; and *the ratios between the "equivalent" quantities of the various species of energy form a single self-consistent set.* This was first shown by the work of Joule who, in the period 1840–1850, established a solid experimental foundation for declaration of the conservation principle. Indeed, one of the first fruits of this work was a demonstration of just the case we have discussed, for Joule was able to show that from a given quantity of mechanical energy one obtains the same amount of heat by either of the routes shown in Fig. 11.

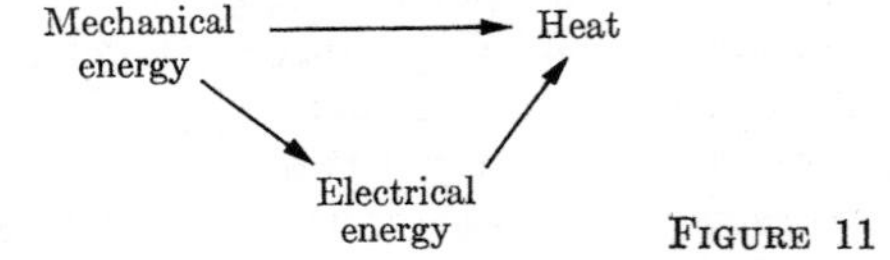

FIGURE 11

Imagine that a certain quantity of heat (q) is received by a system from its surroundings. Suppose that, as a result of this heat input, the system does a certain amount of work (w) on its surroundings. We may well find that the work so delivered by the system is less than the mechanical equivalent of the heat delivered to the system. Accepting the conservation principle, we must then suppose that energy equivalent to the difference $(q - w)$ is somehow stored in the system as an increase of "internal energy" (E), and we write

$$q - w = \Delta E, \qquad q - w = dE. \tag{2}$$

For a finite change   For an infinitesimal change*

For our purposes, either of the above equations is an entirely adequate definition of internal energy. But some may still wish to ask: What *is* internal energy? Classical thermodynamics neither asks nor answers this question. On the kinetic-molecular hypothesis, we may envision internal

* We will use q and w to symbolize, indifferently, both finite and infinitesimal quantities of heat and work respectively.

energy as in part kinetic energy of translational motion—increase in which manifests itself in a rise of temperature. But internal energy must comprise also potential energies that do *not* show themselves in a rise of temperature. Thus a latent heat of vaporization may increase the potential energy of molecules without increasing their kinetic energies, or a heat of decomposition may be stored as the potential energy of atoms or groups of atoms liberated from the bonds that formerly united them, etc. But such speculations do not at all concern classical thermodynamics which, indeed, derives much of its power and generality precisely from its ability to exploit the energy concept while making no more assumption about the "nature" of internal energy than about the constitution of matter.

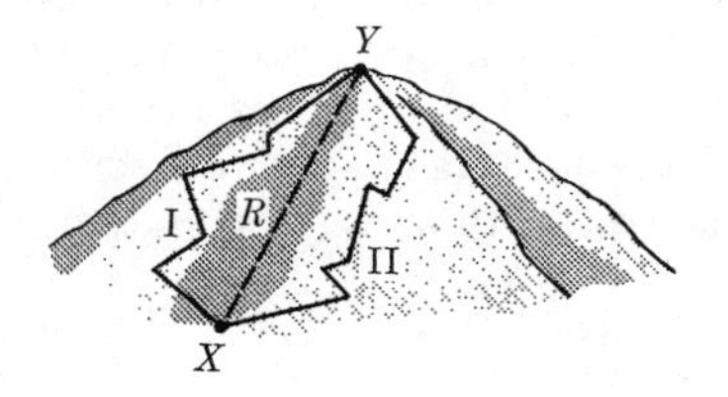

FIGURE 12

Characteristics of a function of state

The change of internal energy associated with the transition from one state to another proves independent of the path taken between them. The situation here is quite analogous to that in many more familiar instances—vertical displacements, for example. Consider that we set out from the base of a mountain, at X, to climb over very rough terrain to its top at Y. Suppose that we can go by either route I or route II, as shown in Fig. 12. Along either route, we will be ascending more or less steeply at some times, traversing horizontally at others, and making longer or shorter descents at still others. In each phase of our journey we can determine (e.g. with an aneroid altimeter) our net vertical ascent or descent *for that phase*. Now, according to which route we choose, we will have two wholly different sets of components of ascent and descent. But, *regardless* of which route we choose, once arrived at Y we can write

$$\begin{aligned}\text{Net vertical displacement} &= \Delta h_{X\to Y}\\ &= \sum \text{Vertical ascents} - \sum \text{Vertical descents.}\end{aligned}$$

That is, the difference of the path components notwithstanding, the net vertical displacement is invariant—independent of the path taken. Note too that when, by a return from Y to the starting point at X, the cycle is completed along any path, e.g., by the more direct aerial-tramway route R, the overall *net* vertical displacement must be nil:

$$\Delta h_{XYX} = \sum \text{Vertical ascents} - \sum \text{Vertical descents} = 0.$$

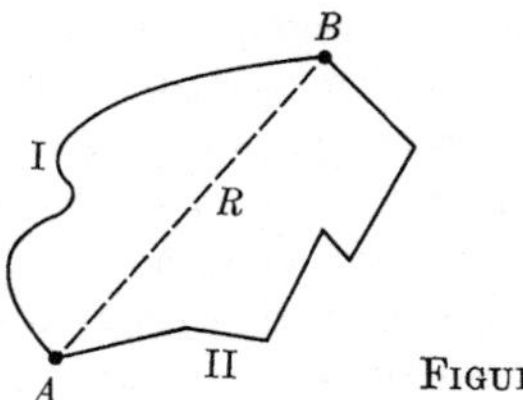

FIGURE 13

Consider now the analogous cycle involving ΔE, q, and w. Of course q and w are not vector quantities, like components of ascent or descent. But analogy still obtains because, as defined by equation (2), ΔE is increased by any positive increment of q and decreased by any positive increment of w. Now from state A of a system, Fig. 13 suggests, passage may be made by either route I or route II to some other state B—whence a return to state A may be made *via* still another route, R. For the complete cycle, restoring the system to its original state, we write

$$\Delta E_{ABA} = \sum q - \sum w = 0.$$

Our justification for writing "$= 0$" is strictly *empirical*. For were this equation false, the ceaseless renewal of the cycle would yield a ceaseless net production or consumption of energy, wholly unlike anything yet encountered in the vast ambit of human experience. Indeed, centuries of failure in the effort to construct perpetual motion machines (and in related efforts) represent the empirical foundation for the energy conservation principle—on the strength of which we have written the last equation.

Now the system can be carried through a complete cycle of changes in either of two ways. When it passes from A to B over route I, and returns *via* R, we write

$$\begin{aligned}\sum q - \sum w &= (q_{\text{I}} + q_R) - (w_{\text{I}} + w_R) = 0,\\ &= (q_{\text{I}} - w_{\text{I}}) + (q_R - w_R) = 0.\end{aligned}$$

When the passage is made over route II, and the return again *via* R,

$$(q_{\text{II}} - w_{\text{II}}) + (q_R - w_R) = 0.$$

Comparison of the last two equations makes it clear that

$$q_{\text{I}} - w_{\text{I}} = q_{\text{II}} - w_{\text{II}} \tag{a}$$

and that, therefore,

$$\Delta E_{\text{I}} = \Delta E_{\text{II}}.$$

Defined as $(q - w)$, ΔE has then—remarkably enough—a property possessed by neither q nor w. For we have seen that both q and w generally vary with the path taken between A and B. As regards w, our findings (seen perhaps with greatest vividness in Figs. 8 and 9) certainly justify the conclusion that ordinarily $w_{\mathrm{I}} \neq w_{\mathrm{II}}$. From equation (a) it then follows also that $q_{\mathrm{I}} \neq q_{\mathrm{II}}$. And yet, though defined in terms of q and w, ΔE is shown by the last equation above to be the *same* for all paths from A to B. Hence ΔE depends *only* on the initial and final states concerned, and we can write

$$\Delta E_{AB} = E_B - E_A.$$

The quantity E is a function of state: we can attribute to each thermodynamic state of a system a characteristic internal energy that is independent of the way that state was arrived at.

How can we establish values for ΔE? As it is defined, so also ΔE is determined—as the difference $(q - w)$. And in one special case such a determination can be reduced to a single measurement. Consider that, save in galvanic cells, chemical reactions are ordinarily run under conditions in which only $P\,dV$ work is possible. Under these conditions a change conducted at *constant volume* shows $dV = 0$, $P_x\,dV = 0$, and $w = 0$. Equation (2) is thus reduced to

$$\Delta E = q_V, \tag{3}$$

where q_V symbolizes the heat transferred when (i) only $P\,dV$ work is possible, and (ii) a constant-volume condition is enforced. How can this equation possibly hold when ΔE is a function of state and q's vary with path? Very simply: the stipulations (i) and (ii) exclude all possible paths save one. And along this particular path q_V has a well-defined value which (like ΔE) depends only on the initial and final states concerned.

To determine the change of internal energy accompanying a reaction brought about inside a rigid sealed ampule, we need only enclose the ampule in, say, an ice calorimeter: the heat transfer measured in these circumstances is q_V, to which we can simply equate ΔE. But such enforcement of constant-volume conditions is usually inconvenient, and sometimes downright impossible. How shall we then proceed?

ENTHALPY

Ordinarily chemical reactions are run not at constant volume but, rather, in systems maintained at *constant pressure* by contact with the air, or with some other gas at atmospheric pressure. More specifically, these reactions most often proceed under a threefold constraint: (i) the external pressure is constant, (ii) the internal pressure is equal to the external pressure, and (iii) only pressure-volume work is possible. In our general statement of the first principle, $\Delta E = q - w$, condition (iii) allows us to substitute for w on the strength of equation (1); and condition (i) then permits the further important simplification indicated as follows:

$$\Delta E = q - \int_{V_1}^{V_2} P_x \, dV$$
$$= q - P_x \int_{V_1}^{V_2} dV = q - P_x(V_2 - V_1) = q - P_x \, \Delta V.$$

In view of condition (ii), we can now replace the external pressure (P_x) by the pressure (P) of the system itself. And at the same time we replace q by q_P—where q_P symbolizes the heat transfer measured under the restrictive conditions (i), (ii), and (iii).

$$\Delta E = q_P - P \, \Delta V. \tag{b}$$

Again symbolizing by the subscripts 1 and 2 the values in initial and final states respectively, we render the Δ terms as follows:

$$E_2 - E_1 = q_P - P(V_2 - V_1),$$

which rearranges to

$$q_P = (E_2 + PV_2) - (E_1 + PV_1).$$

This relation can be given very compact expression in terms of another thermodynamic function. The *enthalpy* (Gr. *enthalpien*, to warm in) is symbolized by H and defined by the equation:

$$H \equiv E + PV. \tag{4}$$

For any change proceeding under restrictions (i) through (iii), the last two equations can be combined, to yield

$$q_P = H_2 - H_1 = \Delta H. \tag{5}$$

The three restrictions eliminate all but one path between terminal states which are, then, the sole determinants of q_P. For the given change, q_P is thus a well-defined quantity—fit to correspond to the alteration of a function of state. And defined as it is by equation (4)—in terms of E, P, and V—enthalpy clearly *is* a function of state. Though lacking the significance of the truly fundamental state functions, energy and entropy, enthalpy is a very *convenient* function. For, by way of equation (5), we can easily determine changes of enthalpy directly from the heat (q_P) transferred under the constant-pressure conditions in which chemical reactions are usually run.

Consider now a change between two states of a system that differ in *both* pressure and volume. For the initial and final states, respectively,

$$H_2 = E_2 + P_2V_2,$$
$$H_1 = E_1 + P_1V_1.$$

Subtracting, we arrive at the following perfectly *general* equation:

$$\Delta H = \Delta E + \Delta(PV). \quad (6)$$

For the important *special* case in which P is stipulated constant, we move the P outside the operator Δ, writing

$$\Delta H = \Delta E + P\,\Delta V. \quad (7)$$

Observe the even simpler one-step derivation of this equation—by direct substitution from equation (5) in equation (b).

Depending on the sign of $\Delta(PV)$ in the general case of equation (6), or the sign of $P\,\Delta V$ in the special case of equation (7), ΔH may be either larger or smaller than ΔE. When gases are involved—and particularly when there is a change (Δn) in the number of moles of gas present—ΔH may differ significantly from ΔE. When *only* gases are involved, and they can be treated as effectively ideal, we can write for *both* the general and special cases:

$$\Delta H = \Delta E + \Delta(nRT) = \Delta E + RT(\Delta n).$$

↖ IF temperature is constant

Only in a constant-volume process can we equate ΔE to the heat transferred (q_V); *only* in a constant-pressure process can we equate ΔH to the heat transferred (q_P). But energy and enthalpy are functions of state, and *both* ΔE and ΔH must have well-defined values for *any* given change of state. The following three examples show how the relations derived in the last two paragraphs can be used to calculate ΔE in a constant-pressure change, and ΔH in a constant-volume change.

▶ *Example 1*

A heat input of 9.71 kcal is required to vaporize 1 mole of water at 1 atm pressure and 100°C(373.15°K). (a) What are ΔH and ΔE for this vaporization process? (b) What are ΔH and ΔE for the inverse process in which one mole of steam condenses at 100°C and 1 atm pressure?

Solution. (a) For this constant-pressure vaporization,

$$\Delta H = q_P = 9.71 \text{ kcal/mole.}$$

Noting that the volume of the liquid vaporized ($V_L \cong 18$ ml) is essentially negligible compared with the volume of the mole of gas produced ($V_G > 25{,}000$ ml), we write

$$P\,\Delta V = P(V_G - V_L) \cong PV_G = RT$$
$$= (2)(373) = 746 \text{ cal/mole} \cong 0.75 \text{ kcal/mole.}$$

Substitution in equation (7) then yields

$$\Delta E = \Delta H - P\,\Delta V = 9.71 - 0.75 = 8.96 \text{ kcal/mole.}$$

(b) In the constant-pressure condensation there is of course an *output* of heat, and here

$$\Delta H = -9.71 \text{ kcal/mole.}$$

Condensation is accompanied by a decrease of volume, so that

$$P\,\Delta V \cong P(-V_G) = -RT = -0.75 \text{ kcal/mole.}$$

By equation (7) we have then

$$\Delta E = \Delta H - P\,\Delta V = -9.71 - (-0.75) = -8.96 \text{ kcal/mole.}$$

Given that E is a function of state, we could of course obtain this last result simply by inspection of answer (a). ◀

▶ *Example 2*

A heat input of 1.435 kcal/mole is required to melt ice at 0°C and 1 atm pressure. Given that in this change 19.63 ml of ice are converted to 18.00 ml of water, determine ΔH and ΔE for the indicated fusion process.

Solution. For this constant-pressure change

$$\Delta H = q_P = 1.435 \text{ kcal/mole.}$$
$$P\,\Delta V = P(18.00 - 19.63) = (1)(-1.63) = -1.63 \text{ ml-atm/mole}$$
$$= -0.00163 \text{ lit-atm/mole} = -0.04 \text{ cal/mole}$$
$$= -0.00004 \text{ kcal/mole.}$$

But with $P\,\Delta V \ll \Delta H$, it follows that

$$\Delta E = \Delta H - P\,\Delta V \cong \Delta H = 1.435 \text{ kcal/mole.}$$ ◀

▶ *Example 3*

At 25°C, 0.816 kcal is released in the detonation of 0.001 mole of solid TNT, $CH_3C_6H_2(NO_2)_3$, in a constant-volume calorimeter. Given that in this reaction 1 mole of TNT yields 5 moles of gaseous N_2 and CO, together with solid and liquid products of negligible volume, calculate ΔH and ΔE per mole of TNT exploded.

Solution. For one mole of TNT in this constant-volume process,

$$\Delta E = q_V = -0.816/0.001 = -816 \text{ kcal/mole.}$$

Neglecting alike the volume of the liquid and solid products and that of the TNT itself, we assume that ΔPV for the reaction arises primarily from the formation of the gaseous CO and N_2.

$$\Delta PV = RT(\Delta n) = (2)(298)(5) = 2980 \text{ cal/mole} \cong 3 \text{ kcal/mole.}$$

And then

$$\Delta H = \Delta E + RT(\Delta n) = -816 + 3 = -813 \text{ kcal/mole.}$$ ◀

When only condensed phases are involved, the solution of Example 2 makes it clear that ordinarily $\Delta H \cong \Delta E$ will be an excellent approximation. Even when there is a change (Δn) in the number of moles of gas present, the answer to Example 3 indicates that if ΔH and ΔE are inordinately large then $\Delta H \cong \Delta E$ may still be a satisfactory approximation (good to better than 0.5% in the TNT example). But significant differences between ΔH and ΔE may emerge when Δn is nonzero *and* ΔH and ΔE are not numerically overpowering: thus in Example 1 the difference amounts to some 8%. Regardless of the size of the differences, the reader will now do well to review the above examples with a view to *explaining* to himself, in words, the relation of ΔH to ΔE in the four possible combinations of (a) exothermic or endothermic change with (b) positive or negative ΔV.

Three further observations are suggested by the above examples. Note first that, unlike the intensive property temperature, heat is an extensive quantity. That is, other things being equal, the amount of heat transferred is directly proportional to the size of the system involved. Calculated as they are from heat measurements, ΔE and ΔH terms are then also extensive quantities in the very same sense. Hence a figure for ΔE or ΔH is meaningful only if we know not only the *identity* but also the *amount* of the change concerned. In the absence of any other indication, a stated ΔE or ΔH should be assumed to refer to a change of one mole of material; and in doubtful cases we will always use barred symbols, like $\Delta\bar{E}$ and $\Delta\bar{H}$, to specify the *molar* change in an extensive variable of state.

A second observation is closely related to the first. Note that, once we have adopted our sign convention for heat transfers, a *positive* sign will automatically attach to ΔE or ΔH terms for an endothermic change, in which heat is transferred *to* the system; and a *negative* sign for an exothermic change, in which heat is transferred *from* the system. Thus no *new* conventional element is required to establish the signs of ΔE and ΔH

terms. But there *is* a new element of convention in the choice of reference state(s) from which H's (or E's) are to be measured—and this brings us to the third and most important point.

Standard states

Suppose we were interested (perhaps as climbers) in the differences in altitude between 100 points on the earth's surface. A list of all these differences involves $(100)(99)/2 = 4950$ entries, but it is easy to be *much* briefer. We need only state the elevation of each of the 100 points above some arbitrary reference level, e.g., "sea level." From these 100 elevations we can at once obtain all 4950 differences of altitude—which, of course, remain wholly unchanged if we set our reference level not at sea level but, say, at the bottom of Death Valley. As we thus handle differences of altitude (and differences of longitude), so also do we handle differences of potential energies, and so also do we handle ΔE's and ΔH's.

Experimentally we cannot measure energies and enthalpies as such, but only the *differences* ΔE and ΔH. Moreover, *only* these differences figure in the statement and use of the principles of thermodynamics. To adopt some arbitrary reference state(s), relative to which energies or enthalpies will be measured, is then purely a matter of convenience—but the gain in convenience is enormous. The elements, in their stablest forms at 25°C and 1 atm pressure, constitute the standard reference states most widely used by chemists.* By convention, an enthalpy of zero is assigned to each of these standard states. That is, for each element in its stablest form at 298.15°K and 1 atm pressure, we write $H^0_{298} \equiv 0$. Here, as everywhere else, the subscript indicates the absolute temperature concerned; and the superscript zero indicates that the term refers to standard conditions, e.g., 1 atm pressure, etc.

THERMOCHEMISTRY AND HESS'S LAW

Because we generally work at constant (atmospheric) pressure, as chemists we are more often concerned with ΔH's than with ΔE's.[12] Sometimes we are anxious to learn the value of ΔH for a change that is difficult to characterize in practice because, for example, it proceeds very slowly; sometimes we wish to know ΔH for a change we are wholly unable to bring about in practice; and always we wish to extract, from a minimum series of measurements, ΔH-values for a maximum series of different changes.

* We need not one but many reference states just because transmutation of the elements is, to say the least, a rarity in purely chemical reactions. For chemists the establishment of a distinct reference state for *each* element is then not merely permissible but desirable.

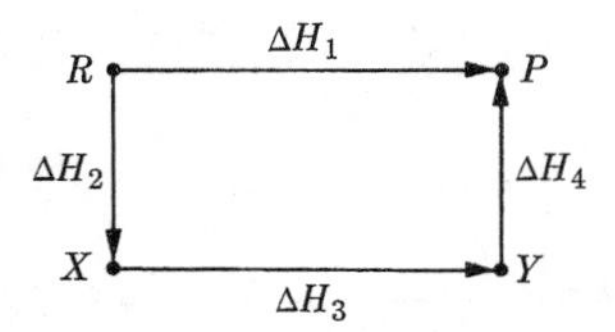

FIGURE 14

Gratification of all these desires is achieved with the aid of Hess's law, the validity of which rests directly on the firm basis of the first principle of thermodynamics.

Suppose we do not know and, perhaps, cannot measure ΔH_1 for the change of some particular reactants (R) to some particular products (P). We can still *calculate* ΔH_1 if we know or can measure the enthalpy changes for the series of reactions $R \rightarrow X$, $X \rightarrow Y$, and $Y \rightarrow P$—as sketched in Fig. 14. Since enthalpy is a function of state, the net change of enthalpy must be exactly the *same* along both routes from R to P, so that

$$\Delta H_1 = \Delta H_2 + \Delta H_3 + \Delta H_4,$$

and ΔH_1 is seen to be readily calculable. The last equation is an expression of Hess's law; the existence of a corresponding equation in ΔE's is self evident.[13]

To illustrate the power of Hess's law, we consider now the reaction

$$\text{C (solid)} + 2\,\text{H}_2 \text{ (gas)} = \text{CH}_4 \text{ (gas)}.$$

Observe how, in order to establish the precise identity of the reaction in question, the physical state of each of the species involved is pinned down by parenthesized indications we shall henceforth abbreviate as s, g, or (for liquids) ℓ. When, as here, no more appears within the parentheses, we will assume that the reaction has been run at 298°K under the standard condition of 1 atm pressure. We properly avoid one residual ambiguity by indicating explicitly that graphite is the particular solid allotrope of carbon that concerns us.

Now though we cannot readily observe the above reaction in actual practice, we can easily establish its ΔH. We have only to bring Hess's law to bear on the enthalpy changes determined for three combustion reactions readily accessible to study:

$$\begin{aligned} \text{C (s, graphite)} + \text{O}_2 \text{ (g)} &= \text{CO}_2 \text{ (g)}, & \Delta H &= -\ \ 94.1 \text{ kcal;} \\ \text{H}_2 \text{ (g)} + \tfrac{1}{2}\,\text{O}_2 \text{ (g)} &= \text{H}_2\text{O } (\ell), & \Delta H &= -\ \ 68.3 \text{ kcal;} \\ \text{CH}_4 \text{ (g)} + 2\,\text{O}_2 \text{ (g)} &= \text{CO}_2 \text{ (g)} + 2\,\text{H}_2\text{O } (\ell), & \Delta H &= -212.8 \text{ kcal.} \end{aligned}$$

We now multiply by 2 the second equation *and the ΔH term appertaining thereto*—for, as an extensive quantity, ΔH is directly proportional to the

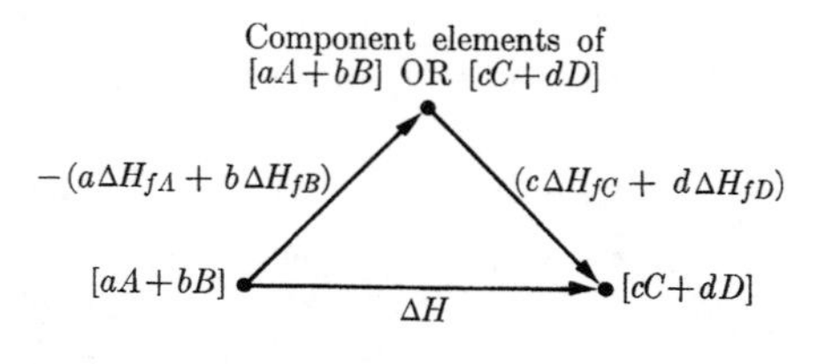

FIGURE 15

quantity of material undergoing the change. Subtraction of the third equation from the sum of the first two then yields

$$\text{C (s, graphite)} + 2\,\text{H}_2\text{ (g)} = \text{CH}_4\text{ (g)},$$
$$\Delta H = -94.1 + 2(-68.3) - (-212.8) \cong -18 \text{ kcal/mole}.$$

We have thus established the molar "heat of formation" of methane from its elements.

Whether determined directly or (as above) indirectly, heats of formation offer occasion for an impressive display of the predictive power of thermodynamics. Given a compilation of heats of formation for each of a number of compounds, we can calculate the enthalpy changes for *all* the myriad chemical reactions that involve only those compounds and their component elements. We illustrate this possibility by subjecting the general reaction $aA + bB = cC + dD$ to the Hess-law style of analysis displayed in Fig. 15. Symbolizing by ΔH_{fA}, ΔH_{fB}, ΔH_{fC}, and ΔH_{fD} the known molar heats of formation of the compounds A, B, C, and D respectively, we seek the enthalpy change (ΔH) attending the overall reaction. We have only to imagine the reaction as carried out in the two steps suggested by the figure. In the first step the a moles of compound A and the b moles of B are decomposed into their component elements. The enthalpy change accompanying *formation* of $[aA + bB]$ *from* their elements is, by definition, $(a\,\Delta H_{fA} + b\,\Delta H_{fB})$; hence $-(a\,\Delta H_{fA} + b\,\Delta H_{fB})$ must represent the enthalpy change in the inverse reaction of *decomposition* of $[aA + bB]$ *into* their component elements. But if our chemical equation has been properly balanced, these must be precisely the elements that yield $[cC + dD]$, in a reaction with enthalpy change $(c\,\Delta H_{fC} + d\,\Delta H_{fD})$. Equating the enthalpy change along the two routes from $[aA + bB]$ to $[cC + dD]$, we write

$$\Delta H = (c\,\Delta H_{fC} + d\,\Delta H_{fD}) - (a\,\Delta H_{fA} + b\,\Delta H_{fB}).$$

A formulation at once briefer and more general can now easily be seen. Let n_R symbolize the number of moles of each reactant (R) consumed; let n_P symbolize the number of moles of each product (P) formed. For the overall reaction, as written, the enthalpy change is then given as

$$\Delta H = \sum n_P\,\Delta H_{fP} - \sum n_R\,\Delta H_{fR}. \tag{c}$$

▶ *Example 4*

For the molar heat of formation of CH_4 we found -18 kcal; and -20 kcal and -22 kcal are cited as the molar heats of formation of gaseous CH_3Cl and HCl respectively. Determine the enthalpy change accompanying the reaction:

$$CH_4\ (g) + Cl_2\ (g) = CH_3Cl\ (g) + HCl\ (g).$$

Solution. We are given heats of formation for only three of the four substances appearing in the equation. But the fourth is the element Cl_2, to which our convention assigns an enthalpy of zero. Substituting then in equation (c) above, we find

$$\Delta H = (-20 - 22) - (-18 + 0) = -24 \text{ kcal.}$$ ◀

If one is prepared to overstep the bounds of thermodynamics, in the region where it borders atomic-molecular theory,[14] one finds within easy reach a predictive device even more strikingly effective than a compilation of heats of formation. But, as always when one leaves the domain of thermodynamics, all gains in predictive power are made at the cost of predictive reliability. This point we illustrate, in a brief digression from strictly thermodynamic considerations, by examining the predictive significance of *bond energies.*

Bond energies

A fundamental physical assumption underlies any tabulation of bond energies. We assume that, between a pair of atoms of any two given elements (A and B), the strength of the A—B bond is wholly independent of the molecular environment in which the atom pair may occur. To break that particular species of bond should then always require the *same* amount of energy—which is, loosely speaking, what we call the "bond energy." Most often bond energies actually refer to hypothetically gaseous species at 0°K, but the (calculable) differences between bond energies at 0°K and bond enthalpies at 298°K are usually negligible compared to what we show on p. 39 to be major shortcomings of the basic physical assumption. Because of their greater immediate usefulness, we shall then speak only of bond enthalpies at 298°K.

How shall we determine the enthalpy change associated with breakage of a given species of bond at 298°K? From the equation

$$C\ (s, \text{graphite}) + 2\,H_2\ (g) = CH_4\ (g), \qquad \Delta H = -18 \text{ kcal},$$

we *cannot* conclude that breakage of a mole of C—H bond is associated with an enthalpy increase of $18/4 = 4.5$ kcal. Such an inference is invalid because, beyond the *formation* of four C—H bonds, the above reaction involves *breakage* of two H—H bonds and, also, *breakage* of whatever bonds link the carbon atom with others of its kind in solid graphite. However, a straightforward application of Hess's law permits us to calculate the C—H bond enthalpy. We know both the heat of dissociation of gaseous hydrogen and the heat of sublimation of solid graphite:

$$\text{H}_2\ (\text{g}) = 2\ \text{H}\ (\text{g}), \qquad \Delta H = 104\ \text{kcal};$$
$$\text{C}\ (\text{s, graphite}) = \text{C}\ (\text{g}), \qquad \Delta H = 171\ \text{kcal}.$$

Multiplying the first equation by 2, and adding it to the second, we find

$$\text{C}\ (\text{s, graphite}) + 2\ \text{H}_2\ (\text{g}) = \text{C}\ (\text{g}) + 4\ \text{H}\ (\text{g}),$$
$$\Delta H = 2(104) + 171 = 379\ \text{kcal}.$$

Subtraction from the first equation in the present paragraph then yields

$$\text{C}\ (\text{g}) + 4\ \text{H}\ (\text{g}) = \text{CH}_4\ (\text{g}), \qquad \Delta H = -18 - 379 = -397\ \text{kcal}.$$

Containing four presumably identical C—H bonds, gaseous methane would thus be broken into its component gaseous atoms by an *average* investment of some 99 kcal per mole of C—H bond broken. For the C—H link, the molar bond enthalpy is then 99 kcal. A similar computation, setting out from the heat of combustion of ethane, leads to

$$2\ \text{C}\ (\text{g}) + 6\ \text{H}\ (\text{g}) = \text{C}_2\text{H}_6\ (\text{g}), \qquad \Delta H = -677\ \text{kcal}.$$

Duly allowing for the $6 \times 99 = 594$ kcal associated with the presence of 6 C—H bonds, we conclude that for the C—C link the bond enthalpy is 83 kcal. And so on. A short tabulation of bond enthalpies at 298°K is given in Appendix IV. Such a tabulation[15] brings within reach a dazzlingly simple calculation of the enthalpy changes accompanying chemical reactions.

▶ *Example 5*

Given that the bond enthalpies for C—H, Cl—Cl, C—Cl, and H—Cl are respectively 99, 58, 78, and 103 kcal/mole, determine ΔH for the reaction: $\text{CH}_4\ (\text{g}) + \text{Cl}_2\ (\text{g}) = \text{CH}_3\text{Cl}\ (\text{g}) + \text{HCl}\ (\text{g})$.

Solution. The reaction involves the following changes:

breaking one C—H bond:	$\Delta H =$	$+99$ kcal (endothermic)
breaking one Cl—Cl bond:	$\Delta H =$	$+58$ kcal (endothermic)
making one C—Cl bond:	$\Delta H =$	-78 kcal (exothermic)
making one H—Cl bond:	$\Delta H =$	-103 kcal (exothermic)
Net:	$\Delta H =$	-24 kcal (exothermic)

To the two significant figures, this is the same result we obtained from the heat-of-formation data. ◀

From equation (c) the enthalpy changes accompanying *all* chemical reactions can be calculated IF we know heats of formation for all the millions of compounds already in hand and yet to be prepared. Used as above, bond enthalpies present a far more alluring prospect. Assume for the moment that all the elements form with each other only single covalent bonds. Knowing enthalpies for only 4950-odd bonds—representing all possible atomic pairings of the 100-odd known elements—we could then calculate the enthalpy changes accompanying all gas-phase reactions. Extension to reactions involving liquids and solids only requires appropriate use of measurable heats of vaporization and heats of fusion. Extension to substances involving double and triple bonds likewise presents no problem in principle: enthalpies for such multiple bonds are easily obtained by the same methods used for single bonds. Using bond enthalpies, we should thus be able to calculate the enthalpy changes associated with *all* chemical reactions, including as yet unknown reactions of compounds as yet unprepared.

Alas, this brilliant promise cannot be fully achieved. The whole bond-enthalpy approach depends, we saw, on the *non*thermodynamic assumption that the strength of the bond formed between a pair of atoms is independent of the molecular environment of those atoms. This assumption proves highly vulnerable. Presumably it is at its best in the series of paraffin hydrocarbons. But even here there are serious failures—such as those that emerge from the calculation demanded by problem 11. For there we find that the three isomeric pentanes—containing in each case four C—C and twelve H—H bonds—have heats of formation that spread over a range of nearly 10%, and that on the average differ by nearly 20% from the heat of formation calculated from bond-enthalpy data. Even more striking failures of bond-enthalpy calculations occur when, in situations of "resonance," spatially extended intramolecular interactions may span far more than two atoms at a time. However attractive in concept, and however useful as a rough guide in practice, the bond-enthalpy method is thus seen to be subject to all the uncertainties characteristic of atomic-molecular hypotheses generally. And on the other hand, however clumsy and demanding in application, the strictly thermodynamic approach based on heats of formation never fails.

HEAT CAPACITY

The specific heat of a substance is the heat input that increases by 1°K the temperature of 1 gm of the substance. The heat capacity (C) is the heat input that increases by 1°K the temperature of 1 mole of the sub-

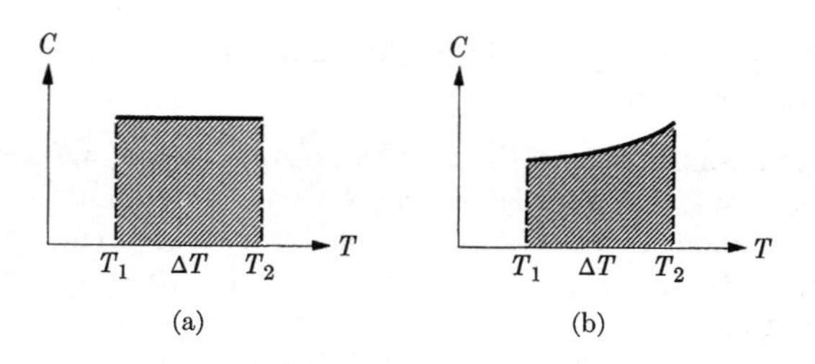

FIGURE 16

stance. Save at comparatively low temperatures, most heat capacities change only very gradually with temperature. If C is effectively constant over the temperature range concerned, the situation is as represented in Fig. 16(a). The shaded area represents the heat input (q) that increases the temperature of one mole of material from T_1 to T_2, and algebraically

$$q = C(T_2 - T_1) = C\,\Delta T.$$

When C is insufficiently approximated as a constant, the situation may appear as in Fig. 16(b). The heat input that increases the temperature from T_1 to T_2 is still represented by the shaded area under the plot of C *vs* T, and this area can always be estimated graphically. However, if we possess an analytical expression for C as $f(T)$, the area is found more easily and accurately by integration:

$$q = \int_{T_1}^{T_2} C\,dT.^*$$

We have spoken of C as though it were a well-defined quantity, but of course the heat required to pass from one specified state of a system to another will vary with the path taken. The heat capacity can then *become* a well-defined quantity only by virtue of a specification of path. Actually, we find useful *two* species of heat capacities, corresponding to *two* simply specified paths—that at constant volume and that at constant pressure.

At constant volume, over a temperature range in which the heat capacity is constant, the first equation of this section becomes

$$q_V = C_V\,\Delta T.$$

* For many substances C can be expressed by an empirical power series of the type $C = a + bT + cT^{-2}$. In such a case, for one mole of material, we will write

$$q = \int_{T_1}^{T_2} (a + bT + cT^{-2})\,dT = a(T_2 - T_1) + \frac{b}{2}(T_2^2 - T_1^2) - c\left(\frac{1}{T_2} - \frac{1}{T_1}\right).$$

But, in view of equation (3), we can then at once write

$$\Delta E = C_V \, \Delta T.$$

If, over the temperature range concerned, C_V is *not* constant, substitution in the second equation of this section yields

$$\Delta E = q_V = \int_{T_1}^{T_2} C_V \, dT.$$

All of these equations have been written for one mole of material, but the corresponding expressions for n moles are readily seen to be

$$\Delta E = n\int_{T_1}^{T_2} C_V \, dT \underset{\text{IF } C_V \text{ is constant}}{=} nC_V(T_2 - T_1) = nC_V \, \Delta T. \quad (8)$$

At constant pressure analogous expressions can be found by inspection. For one mole of material of constant heat capacity we have first

$$q_P = C_P \, \Delta T.$$

And, in view of equation (5),

$$\Delta H = C_P \, \Delta T.$$

If C_P varies over the temperature range concerned, then, substituting as before, we find

$$\Delta H = q_P = \int_{T_1}^{T} C_P \, dT.$$

Lastly, the general expression for n moles of material assumes this form:

$$\Delta H = n\int_{T_1}^{T_2} C_P \, dT \underset{\text{IF } C_P \text{ is constant}}{=} nC_P(T_2 - T_1) = nC_P \, \Delta T. \quad (9)$$

Ideal gases

For an ideal gas, the heat capacities C_P and C_V are linked by an important relation we can easily derive as soon as we have fully defined what we mean by an ideal gas.[16] Apart from the equation $PV = RT$, the other important defining characteristic was first established experimentally by Joule, in the classic experiment sketched in Fig. 17. The system consists of a vessel with two bulbs: one evacuated, the other containing the gas to be examined. This system is allowed to come to temperature equilibrium in a well-insulated water bath containing a thermometer. When the stopcock is opened, the gas, originally confined in volume V_1, distributes

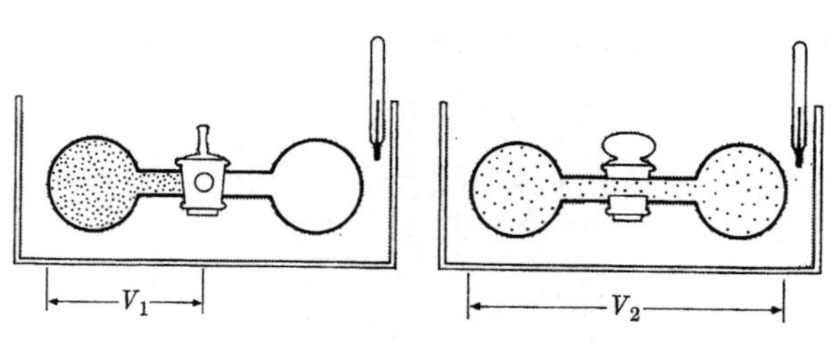

FIGURE 17

itself over the entire volume V_2. From the unchanged reading of the thermometer, Joule inferred that this abrupt expansion into a vacuum is accompanied by *zero* net heat transfer to or from the system. Using more sensitive modern equipment, we find that with real gases (characterized by small but finite intermolecular attractions) the experiment actually indicates very small but finite values for q. However, these values become smaller the lower the initial pressure of the real gas, and approach zero in the limit $P \to 0$ where we suppose a real gas behaves ideally. And so we have come to make it part of our definition of the ideal gas that it must yield $q = 0$ in this experiment.

What is the thermodynamic significance of this behavior? When the gas expands into the vacuum, *no* work is performed by the system on its surroundings. For the isothermal expansion of an ideal gas into a vacuum, we must then write *both* $q = 0$ and $w = 0$. But the first principle of thermodynamics now demands that $\Delta E = q - w = 0$. That is, the isothermal expansion of an ideal gas into a vacuum is characterized by *zero* change in internal energy. At first sight this seems a very strange conclusion. To be sure, if we imagine that the internal energy of the gas inheres in molecular motions that remain as energetic in a large as in a small volume, then indeed the expansion *should* yield $\Delta E = 0$. Yet, on the other hand, the expanded gas can no longer supply all the work recoverable from the compressed gas, and we feel that the expanded gas must *somehow* differ from the compressed gas. That feeling is perfectly sound—and in the next chapter we will find a way to express the difference—but the difference is *not* one of internal energy.

In the peculiarly simple case of expansion into a vacuum, with $q = w = 0$, we have found that, at constant temperature, an ideal gas has the same internal energy when it occupies the volume V_2 as when it occupies the volume V_1. Were the isothermal expansion from V_1 to V_2 to proceed in some different manner, e.g., against some finite external pressure P_x, the values of q and w might well differ from zero. But, given that E is a function of state, we may be certain that if $\Delta E = 0$ when the gas passes from V_1 to V_2 in the one way, then $\Delta E = 0$ when the gas

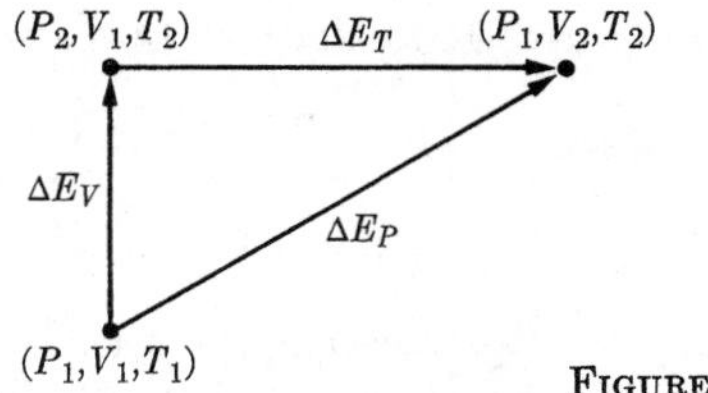

FIGURE 18

passes from V_1 to V_2 in any other way. Considering that V_1 and V_2 can represent any two volumes, we thus arrive at an enormous generalization of our first finding. For the change of internal energy in *every* isothermal expansion or compression of an ideal gas, we can invariably write $\Delta E = 0$.

Consider now the heat capacities displayed by one mole of ideal gas, standing initially at pressure P_1, volume V_1, and temperature T_1. Figure 18 shows two routes by which the mole of gas can be brought to a final state characterized by (P_1, V_2, T_2), where we choose $T_2 = T_1 + 1°\text{K}$. Along one route, the gas is first heated at constant volume to the state (P_2, V_1, T_2), and then expanded at constant temperature to the state (P_1, V_2, T_2). As indicated in the figure, we symbolize by ΔE_V and ΔE_T respectively the energy changes in the constant-volume and constant-temperature steps. Along the second route, the gas is heated at constant pressure to the same final state (P_1, V_2, T_2), and we symbolize by ΔE_P the energy change along this one-step route. Now by Hess's law

$$\Delta E_P = \Delta E_V + \Delta E_T.$$

For the constant-volume heating of one mole of gas, from T_1 to $T_2 = T_1 + 1$, equation (8) gives

$$\Delta E_V = q_V = C_V(T_2 - T_1) = C_V.$$

In the subsequent isothermal expansion from V_1 to V_2 we have just seen that for an ideal gas we can write

$$\Delta E_T = 0.$$

Finally, for the constant-pressure heating of the gas from T_1 to T_2, equations (6) and (9) together yield:

$$\Delta E_P + \Delta PV = \Delta H = q_P = C_P(T_2 - T_1) = C_P.$$

For the one mole of ideal gas, $\Delta PV = \Delta RT = R(T_2 - T_1) = R$. Therefore

$$\Delta E_P = C_P - R.$$

Substituting in the Hess's-law equation above, we find

$$C_P - R = C_V + 0$$

or

$$C_P = C_V + R. \tag{10}$$

This relation expresses the sole empirical basis that J. R. Mayer could find, 130 years ago, for his argument that work and heat are both species of a general "energy" that is conserved. Given the values of C_P and C_V for air, measured by Gay-Lussac and others, Mayer was able to estimate that one calorie of heat is equivalent to 3.6 joules of mechanical work (modern value, $\sim$4.2 joules).

Equation (10) has been abundantly confirmed for real gases at reasonably low pressures, and the heat-capacity ratio

$$\gamma \equiv \frac{C_P}{C_V} = \frac{C_V + R}{C_V}$$

offers an interesting criterion for distinguishing monatomic and polyatomic gases. Again somewhat overstepping the bounds of classical thermodynamics, let us examine how this comes about. In deriving the ideal-gas law from the kinetic-molecular theory, one of the important way-stops is the equation

$$PV = \tfrac{1}{3}Nm\overline{u}^2.$$

Here the N gas particles—each with mass m and (root-mean-square) average speed $\overline{u}$—are supposed to produce the pressure P by their bombardment of the walls of the container with volume V. Multiplying both sides of the last equation by $-3/2$, we find

$$N(\tfrac{1}{2}m\overline{u}^2) = \tfrac{3}{2}PV.$$

For one mole of gas, N is Avogadro's number, and the expression on the left simply represents the total translational kinetic energy ($\overline{E}_{\text{trans}}$) of all the particles present. Substituting on the right from the ideal-gas law $PV = RT$, we arrive at

$$\overline{E}_{\text{trans}} = \tfrac{3}{2}RT.$$

Suppose that one mole of a *monatomic* ideal gas is heated at constant volume through some short temperature interval ΔT. We see no way in which the added energy can be accommodated save as an increment in $\overline{E}_{\text{trans}}$. That is, for this monatomic gas, the *entire* change in internal energy should be expressible as: $\Delta\overline{E} = \Delta\overline{E}_{\text{trans}} = \tfrac{3}{2}R\,\Delta T$. Substitution for ΔE from equation (8) then yields

$$C_V\,\Delta T = \tfrac{3}{2}R\,\Delta T \qquad \text{or} \qquad C_V = \tfrac{3}{2}R.$$

TABLE 1

HEAT-CAPACITY RATIO OF GASES AT A PRESSURE OF 1 ATM
(T_{BP} = boiling-point temperature)

Elements	γ
He, Ne, A (near 25°C)	1.67
Hg (near T_{BP} = 356°C)	1.67
Na (near T_{BP} = 892°C)	1.68
H_2, O_2, N_2 (near 25°C)	1.40
H_2O (near T_{BP} = 100°C)	1.32

On a proposition of the kinetic-molecular theory, we have thus constructed an *a priori* calculation of C_V for an ideal monatomic gas. And with $C_V = \frac{3}{2}R$, equation (10) now requires that $C_P = \frac{5}{2}R$, so that $\gamma = 1.67$. This result was actually predicted almost a decade before an experimental measurement was finally achieved in 1876, for mercury vapor, and the *general* excellence of the agreement with experiment is readily seen from Table 1.

In constant-volume heating of the ideal monatomic gas, all the energy input goes into an increase of the translational kinetic energy. With polyatomic gases, on the other hand, part of the energy input is diverted into increasing the energies of rotational and vibrational motions—"degrees of freedom" wholly absent in monatomic gases. The occurrence of this diversion means that, for polyatomic gases, a heat input *greater* than $\frac{3}{2}R$ is required to bring about the increase of $\overline{E}_{\text{trans}}$ that corresponds to a 1°K rise of temperature. And the larger the number of degrees of freedom over which the added energy must be spread, the greater will be the margin by which C_V must exceed $\frac{3}{2}R$, and the smaller will be the ratio $\gamma = C_P/C_V = 1 + (R/C_V)$. Now all but the monatomic gases have such "extra" degrees of freedom fully operative at and above room temperature. For recognition of a monatomic gas, $\gamma = 1.67$ thus becomes a highly distinctive criterion which, indeed, played an important role in the earliest characterization of the noble gases discovered at the close of the 19th century.

KIRCHHOFF'S EQUATIONS

Consider the general reaction

$$aA + bB = lL + mM.$$

Suppose we have measured ΔH_1, the change of enthalpy for the reaction at some temperature T_1. There is then no need actually to measure ΔH at any other temperature IF we possess C_P-values for the various reactants and products involved. Given such values—$(C_P)_A$, $(C_P)_B$, $(C_P)_L$, $(C_P)_M$—

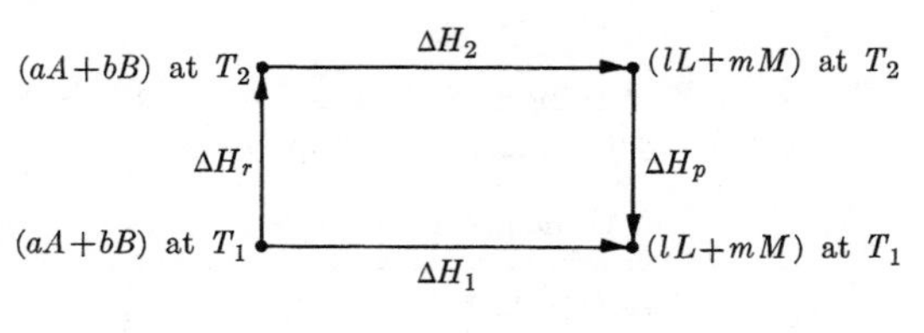

FIGURE 19

we can *calculate* ΔH_2 at any other temperature T_2, by carrying out the analysis suggested by the layout shown in Fig. 19. Enthalpy being a function of state, the enthalpy change must be the same for both the indicated paths from reactants at T_1 to products at T_1. Therefore:

$$\Delta H_1 = \Delta H_r + \Delta H_2 + \Delta H_p.$$

For the constant-pressure heating of the reactants from T_1 to T_2, we write

$$\Delta H_r = \int_{T_1}^{T_2} [a(C_P)_A + b(C_P)_B] \, dT.$$

Similarly, to return the products from T_2 to T_1, we must have

$$\Delta H_p = \int_{T_2}^{T_1} [l(C_P)_L + m(C_P)_M] \, dT.$$

Substituting these values and rearranging terms, we find

$$\Delta H_2 = \Delta H_1 - \int_{T_1}^{T_2} [a(C_P)_A + b(C_P)_B] \, dT - \int_{T_2}^{T_1} [l(C_P)_L + m(C_P)_M] \, dT.$$

By inverting the limits of the second integral, with a corresponding change of its sign, we can combine terms to get

$$\Delta H_2 = \Delta H_1 + \int_{T_1}^{T_2} \{[l(C_P)_L + m(C_P)_M] - [a(C_P)_A + b(C_P)_B]\} \, dT.$$

The difference inside the braces represents the quantity of heat required to increase by 1°K the temperature of all the products collectively, *less* the quantity of heat that produces the same rise of temperature in the corresponding group of reactants. Let us denote by ΔC_P this overall change in the heat capacity of the system, consequent to the reaction. Then:

$$\Delta H_2 = \Delta H_1 + \int_{T_1}^{T_2} \Delta C_P \, dT. \tag{11}$$

This is Kirchhoff's equation for a change at constant pressure; for a change at constant volume, the corresponding equation in ΔE and ΔC_V can be set down by inspection.

To calculate ΔH_2 from a measurement of ΔH_1, we must still evaluate the integral $\int_{T_1}^{T_2} \Delta C_P \, dT$. How shall we do so? Following Kopp (1864) we find that, at any temperature well removed from 0°K, a given element has much the same heat capacity when bound in *any* of its solid compounds whatsoever and, if nongaseous, even when free. Were Kopp's law perfectly accurate and universally applicable, ΔC_P would always be zero. Because the law falls short of perfect accuracy and generality, ΔC_P's are finite, though often negligibly small. In that event, by setting $\Delta C_P = 0$ we can entirely suppress the integral, and we then treat ΔH as effectively constant over at least short ranges of temperature. Good enough for most of the cases that will concern us, this simple-minded approach is insufficient when the span of temperature is broad and/or we seek a result of high quality. As the next simplest policy, we may treat ΔC_P not as zero but as a constant—in which case the last equation becomes

$$\Delta H_2 = \Delta H_1 + \Delta C_P(T_2 - T_1). \tag{12}$$

Securing a figure for ΔC_P by actually adding and subtracting respectively the heat capacities of the products and reactants—as they are tabulated for some *one* temperature, say T_1—we can solve the last equation for ΔH_2. One need proceed more elaborately only in cases far more demanding than any we will face.*

▶ *Example 6*

The standard heat of formation listed for gaseous NH_3 is -11.02 kcal/mole at 298°K. Given that at 298°K the constant-pressure heat capacities of gaseous N_2, H_2, and NH_3 are respectively 6.96, 6.89, and 8.38 cal/mole-°K, determine ΔH^0_{298} and ΔH^0_{773} for the reaction

$$\tfrac{1}{2} N_2 \text{ (g)} + \tfrac{3}{2} H_2 \text{ (g)} = NH_3 \text{ (g)}.$$

Solution. Having adopted a convention that assigns zero as the enthalpy of the elements in their standard states at 298°K, we can at once write

$$\Delta H^0_{298} = \Delta H^0_{f\,NH_3} - 0 - 0 = -11.02 \text{ kcal/mole}.$$

* Expressing C_P, for each product and each reactant, as a power series of the form $C_P = a + bT + cT^{-2}$, one then adds and subtracts to find for ΔC_P an expression of the form $\Delta C_P = \Delta a + \Delta bT + \Delta cT^{-2}$. Substituting this in equation (11), and actually performing the integration, one obtains

$$\Delta H_2 - \Delta H_1 = \Delta a(T_2 - T_1) + \frac{\Delta b}{2}(T_2^2 - T_1^2) - \Delta c\left(\frac{1}{T_2} - \frac{1}{T_1}\right).$$

For the reaction that concerns us, at 298°K we have

$$\Delta C_P = 8.38 - \tfrac{1}{2}(6.96) - \tfrac{3}{2}(6.89) = -5.43 \text{ cal/mole-°K}.$$

Over the 475° our calculation of ΔH^0_{773} must span, this ΔC_P is much too large to be ignored. Let us however venture to treat as constant, over the entire range 298–773°K, the value of ΔC_P we have established for 298°K only. By equation (12),

$$\Delta H^0_{773} = \Delta H^0_{298} + \Delta C_P(773 - 298)$$
$$= -11{,}020 + (-5.43)(475) = -13{,}600 \text{ cal/mole.}$$ ◀

Explosions and flames[17]

Given that ΔH and ΔE *do* change very significantly over extended ranges of temperature, the calculation of the peak temperature reached in an explosion may seem to pose an exceedingly difficult problem. But this problem has an amusingly, and significantly, simple solution—which can be applied also in the determination of the maximum temperature reached in a flame, and so on. For concreteness, consider a reaction in which the gases X and Y are brought together in the correct proportion to form gaseous Z by the strongly exothermic reaction: $X + Y = Z$. Suppose that we are given a figure for $(C_V)_Z$ and, further, that the reaction is to be initiated at a temperature (T_i) for which we know the molar energy change ($\Delta\overline{E}_i$) accompanying the formation of Z by the indicated reaction. We are now asked for the peak temperatures reached when this reaction is run in two different ways: (I) at constant volume, in a sealed bomb to the walls of which we may assume no heat is lost; and (II) at constant pressure, in a flame formed by mixing streams of X and Y, with the heated gas doing work by expanding reversibly against the surrounding atmosphere but, we will assume, losing no heat to the surroundings.

Even given the simplifying assumptions of no heat loss—and many more data—the above problems look pretty hopeless. Consider that in (I) the final temperature will be a function of a heat of reaction that changes progressively as the temperature rises with the advance of the reaction. Consider further that the temperature rise actually produced by any particular fractional advance of the reaction will depend on the heat capacity of the particular gas mixture then present, which will in turn depend on the extent to which the reaction has already proceeded. Part (II) involves both these complications, and poses in addition the difficulty of estimating pressure-volume work, the magnitude of which depends on the final volume, and thus involves the (unknown) final temperature. And so on and on. From this exceedingly grim-looking set of interlocked variables we can, however, escape very easily—if we have but the wit to substitute, for the actual path of the reaction, another more readily analyzed.

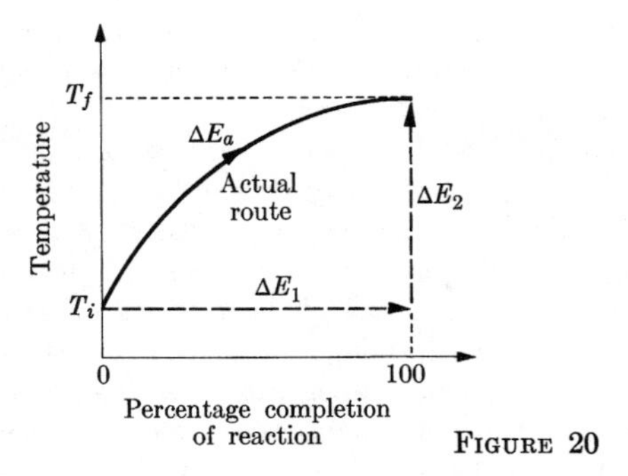

FIGURE 20

This is an expedient open to us whenever, as here, we can set up our problem in terms of variables of state.

Instead of trying to follow the complex evolution of the actual reaction, we conceive it resolved into the two hypothetical steps indicated in Fig. 20. In the first step we imagine the reaction $X + Y = Z$ carried to completion at constant volume, but with strong enough cooling to hold the temperature steady at the original value T_i. In the second step we imagine the reaction product (Z) heated, at constant volume, from T_i to whatever may be the final temperature T_f. Each of these steps is tractable and, since together they connect the initial and final states characteristic of the actual reaction, we have

$$\Delta E_a = \Delta E_1 + \Delta E_2.$$

Here ΔE_a symbolizes the energy change along the actual route of the reaction. And we already know all we need to know about ΔE_a. For this constant-volume reaction, proceeding with no loss of heat to the walls of the bomb, equation (3) yields $\Delta E_a = q_V = 0$. Consequently,

$$\Delta E_1 + \Delta E_2 = 0.$$

What about ΔE_1, for that first step in which the reaction is run to completion at the original temperature? What ΔE_1 represents is clearly the product $n\,\Delta \overline{E}_i$, where n is the number of moles of Z formed in the reaction having the given molar energy change $\Delta \overline{E}_i$ at temperature T_i. What about ΔE_2 for that second step in which the n moles of Z are heated from the initial to the final temperature? Assuming that $(C_V)_Z$ is effectively constant over the range T_i to T_f, we can call on our equation (8) to write $\Delta E_2 = q_V = n(C_V)_Z(T_f - T_i)$. Substituting these values for ΔE_1 and ΔE_2 in the last equation above, we find

$$n\,\Delta \overline{E}_i + n(C_V)_Z(T_f - T_i) = 0, \qquad \text{or} \qquad (T_f - T_i) = \frac{-\Delta \overline{E}_i}{(C_V)_Z}.$$

The minus sign attaching to $\Delta\overline{E}_i$ is quite in order: in an exothermic reaction $\Delta\overline{E}_i$ is negative, and from $-\Delta\overline{E}_i$ we duly obtain the positive temperature rise expected in such a reaction.

In the last equation we have found an extraordinarily simple solution to what seemed an extraordinarily difficult problem. But when dealing with strongly exothermic reactions, we must take care not to *over*simplify. For one thing, $(C_V)_Z$ is unlikely to be constant over the great span of temperature then involved, in which case we can proceed only by expressing $(C_V)_Z$ as $f(T)$ and performing the analysis demanded in problem 15. Furthermore, if at the peak temperature Z is very extensively dissociated into X and Y, this will make nonsense of our tacit assumption that the reaction goes to completion. Though our simple solution is perfectly sound in principle, we must exercise *some* discrimination in applying it in practice.

Given the above solution to part (I), part (II) of the problem presents no new difficulties. For our analysis of the constant-pressure process we do best to use $\Delta\overline{H}_i$ and $(C_P)_Z$ rather than $\Delta\overline{E}_i$ and $(C_V)_Z$. We call on equation (6) to calculate $\Delta\overline{H}_i$ from the given $\Delta\overline{E}_i$, and on equation (10) to calculate $(C_P)_Z$ from the given $(C_V)_Z$. The analysis then follows an almost identical course. Much as before, we can write

$$\Delta H_a = \Delta H_1 + \Delta H_2.$$

With no heat loss attendant upon this constant-pressure process, equation (5) gives us $\Delta H_a = q_P = 0$. Appropriate substitutions for ΔH_1 and ΔH_2 then yield for the constant-pressure flame,

$$(T_f - T_i) = \frac{-\Delta\overline{H}_i}{(C_P)_Z}.$$

The performance of problems 17–19 will show how easily these important equations for $(T_f - T_i)$ can be adjusted to apply to reactions other than $X + Y = Z$.

ADIABATIC PROCESSES

The simplifying assumption of *no* heat loss permitted us to treat explosions and flames as "adiabatic" processes, i.e., processes for which we can by definition write $q = 0$. Many fast processes can be approximated as adiabatic, simply because they are completed before there is time for any substantial heat transfer. The approximation is useful because adiabatic processes are often relatively easy to analyze. Why? Recall how, by imposing a constant-volume restriction on a system capable of pressure-volume work *only*, we were able to reduce the first principle to the very simple relation $\Delta E = q$. Just so, imposition of an adiabatic restriction on *any* system reduces the first principle to another very simple relation of only two terms:

$$-\Delta E = w. \tag{13}$$

Reasonably enough, this equation asserts that work delivered by an adiabatic system is done at the expense of its own internal energy.

Adiabatic expansion of an ideal gas

How shall we apply equation (13) to the adiabatic expansion of an ideal gas? We already know how to express w as $\int P_x\,dV$. But it is *not* so obvious how we are to express ΔE for an adiabatic expansion in which neither pressure nor volume nor temperature is constrained to constancy. However, as always when we deal with a function of state, we can determine the overall ΔE by summing partial ΔE's readily evaluated for a suitably chosen series of steps that lead to the same final state.

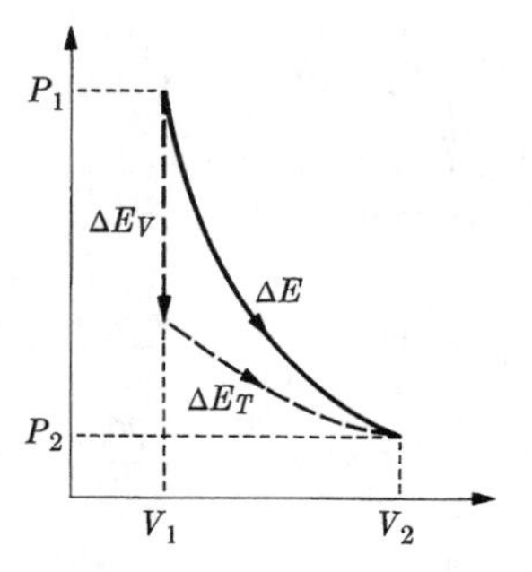

FIGURE 21

Figure 21 displays a two-step route between the initial and final states linked directly in the adiabatic expansion of an ideal gas. For the first step, ΔE_V symbolizes the energy change associated with constant-volume cooling of the gas to its final temperature (T_2). For the second step, ΔE_T symbolizes the energy change associated with the isothermal expansion of the gas to its final volume (V_2). For the overall process, the energy change (ΔE) is then expressible as

$$\Delta E = \Delta E_V + \Delta E_T.$$

But we found earlier that the isothermal expansion of an ideal gas proceeds with *nil* change of internal energy. Hence $\Delta E_T = 0$. And equation (8) yields

$$\Delta E_V = q_V = n\int_{T_1}^{T_2} C_V\,dT.$$

On substituting for both ΔE_T and ΔE_V in the equation just above, we find

$$\Delta E = n\int_{T_1}^{T_2} C_V\,dT + 0 = n\int_{T_1}^{T_2} C_V\,dT.$$

Now we can substitute for both ΔE and w in equation (13), to obtain the following expression applicable to *all* adiabatic expansions and compressions of an ideal gas:

$$-\Delta E = -n\int_{T_1}^{T_2} C_V\, dT = \int_{V_1}^{V_2} P_x\, dV = w. \tag{14}$$

▶ *Example 7*

Standing initially at 298°K and a pressure of 10 atm, an ideal monatomic gas (with $C_V = \frac{3}{2}R$) expands adiabatically and irreversibly against a constant external pressure of 1 atm. What is the temperature of the gas when its own pressure has at last dropped to 1 atm?

Solution. With both P_x and C_V constant, the integrals in equation (14) are easily eliminated, leaving

$$-\Delta E = -nC_V(T_2 - T_1) = P_x(V_2 - V_1) = w.$$

We know neither the number of moles nor the volumes, but these all vanish when we make judicious substitutions on the strength of the ideal-gas law:

$$-n(\tfrac{3}{2}R)(T_2 - 298) = (1)\left(\frac{nRT_2}{P_2} - \frac{nRT_1}{P_1}\right),$$

$$-\tfrac{3}{2}(T_2 - 298) = \frac{T_2}{1} - \frac{298}{10},$$

which yields for the final temperature $T_2 = 191$°K. ◀

Equation (14) applies to *both* reversible and irreversible adiabatic expansions of ideal gases. But the reversible expansions are most readily handled not with equation (14) itself but, rather, with other relations we now derive from (14). In a reversible expansion, we recall from Chapter 1, P_x differs only infinitesimally from, and can be replaced by, the pressure (P) of the gas inside the system. Thus we come to rewrite equation (14) as

$$-nC_V\, dT = P\, dV = \frac{nRT}{V}\, dV,$$

$$\frac{dT}{T} = -\frac{R}{C_V}\cdot\frac{dV}{V}.$$

If, like R, C_V is constant, the integration is easy:

$$\ln\frac{T_2}{T_1} = -\frac{R}{C_V}\ln\frac{V_2}{V_1} = \frac{R}{C_V}\ln\frac{V_1}{V_2} = \ln\left[\frac{V_1}{V_2}\right]^{R/C_V}, \tag{d}$$

so that

$$\frac{T_2}{T_1} = \left[\frac{V_1}{V_2}\right]^{R/C_V}, \tag{e}$$

whence $$T(V)^{R/C_V} = \text{constant}. \tag{15}$$

We obtain another useful relation by using the ideal-gas law for substitutions on the left side of equation (e), while we also reconstruct the exponent on its right side with the aid of equation (10).

$$\frac{P_2V_2}{P_1V_1} = \left[\frac{V_1}{V_2}\right]^{R/C_V} = \left[\frac{V_1}{V_2}\right]^{(C_P-C_V)/C_V} = \left[\frac{V_1}{V_2}\right]^{(\gamma-1)},$$

whence $$P(V)^{\gamma} = \text{constant}. \tag{16}$$

Bear in mind that, like *all* the other equations written in this paragraph, this elegantly simple expression applies only to the *reversible adiabatic* expansion or compression of an *ideal gas.*

▶ *Example 8*

Standing initially at a pressure of 10 atm in a 1 liter container, 0.410 mole of ideal monatomic gas (with $\gamma = \frac{5}{3}$) expands adiabatically and reversibly until its pressure falls to 1 atm. What is its final temperature and volume, and how much work is done in the expansion?

Solution. Substitution in (16) at once yields

$$(10)(1)^{5/3} = (1)(V_2)^{5/3}.$$

Cubing both sides of the equation, and then extracting the fifth root of 10^3, we find $V_2 = 4$ liters. Then

$$T_2 = \frac{P_2V_2}{nR} = \frac{(1)(4)}{(0.410)(0.082)} = 119°\text{K}.$$

The same kind of ideal-gas-law calculation yields as the original temperature $T_1 = 298°\text{K}$. With

$$C_V = \tfrac{3}{2}R \cong 3 \text{ cal/mole-°K} \cong 12.5 \text{ joule/mole-°K},$$

the monatomic gas does an amount of work now easily calculable as

$$w = -\Delta E = -nC_V(T_2 - T_1) = -0.410(12.5)(119 - 298) = 920 \text{ joules}.$$

Contrast of adiabatic and isothermal expansions of an ideal gas

From the outcome of the Joule experiment we inferred that $\Delta E = 0$ in the isothermal expansion or compression of an ideal gas. Joining this finding with the statement of the first principle given in equation (2), we deduce that, in *all isothermal* expansions or compressions of an *ideal gas,*

$$q = w.$$

This result permits us to extend two relations first written in Chapter 1.

As a *general* expression for pressure-volume work, we there wrote

$$w = \int_{V_1}^{V_2} P_x\,dV \underset{\text{IF } P_x \text{ is constant}}{=} P_x(V_2 - V_1) = P_x\,\Delta V. \qquad (1)$$

For any *isothermal* expansion or compression of an *ideal gas,* we can now rewrite equation (1) as

$$q = w = \int_{V_1}^{V_2} P_x\,dV \underset{\text{IF } P_x \text{ is constant}}{=} P_x(V_2 - V_1) = P_x\,\Delta V. \qquad (17)$$

By integration of equation (1), for the very special case of a *reversible isothermal* expansion or compression of an *ideal gas,* in Chapter 1 we obtained an equation (c) reading:

$$w_{\text{rev}} = nRT \ln \frac{V_2}{V_1} = 2.30\,nRT \log \frac{V_2}{V_1} = 2.30\,nRT \log \frac{P_1}{P_2}.$$

This we can now rewrite as

$$q_{\text{rev}} = w_{\text{rev}} = nRT \ln \frac{V_2}{V_1} = 2.30\,nRT \log \frac{V_2}{V_1} = 2.30\,nRT \log \frac{P_1}{P_2}, \qquad (18)$$

where q_{rev} represents the heat transfer associated with w_{rev} in the reversibly conducted change. Thus when 1 liter of gas at a pressure of 10 atm expands isothermally and reversibly until its pressure drops to 1 atm, the calculation given on p. 19 is seen now to determine *both* w_{rev} and q_{rev}:

$$q_{\text{rev}} = w_{\text{rev}} = 2.30(10) \log \tfrac{10}{1} = 23 \text{ lit-atm} = 2330 \text{ joules}.$$

Observe that the work output in this reversible isothermal expansion is more than twice the work output in the corresponding reversible adiabatic expansion just examined in Example 8. There also 1 liter of ideal gas at a pressure of 10 atm expanded reversibly until its pressure fell to 1 atm, but the final volume then reached was *not* the 10 liters found in the isothermal expansion but only 4 liters. All this is readily understandable. In the adiabatic expansion, with $q = 0$ by definition, all work is done at the expense of the internal energy of the gas: as it does work in expanding, its temperature must then decline, and its pressure falls off in response to *both* the increase in volume and the decrease in temperature. In the isothermal expansion, on the other hand, the temperature is stabilized by an inward flow of heat that exactly matches the work output; and the slower decline of pressure then permits the gas to expand to a greater volume and to deliver a greater output of work. Figure 22 displays graph-

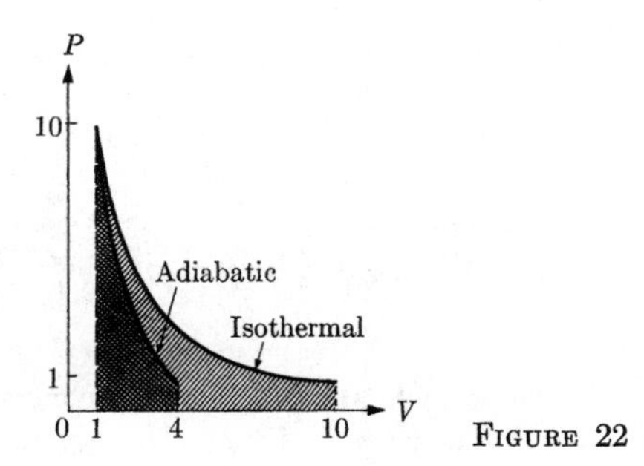

FIGURE 22

ically the relation of the reversible isothermal and adiabatic expansions under discussion. That the adiabatic line always slopes more steeply than the corresponding isotherm is, for the ideal gas, a proposition easily demonstrable in the manner suggested in problem 22—from which we can thus draw reassuring support for a tacit assumption already introduced in sketching Fig. 21.

Characterized by the relation $q = w$, the isothermal expansion of an ideal gas seems to represent a process remarkably adapted for the conversion of heat into work with a theoretical maximum efficiency of 100%. That this beguiling possibility is actually wholly chimerical will emerge as, in the next chapter, we undertake a general study of the efficiency of heat engines.

3

The Second Principle of Thermodynamics

However drastic it may appear, in every change we surmise a "something" that remains constant. From the very beginning of the modern era, some men (e.g., Descartes) have conceived that "something" as more or less close kin to what we would call energy. And energy—or, better, mass-energy—*is* surely conceived by us as a "something constant" enduring through all change. The first principle of thermodynamics thus gives quantitative expression to our firm conviction that "plus ça change, plus c'est la même chose."

We have another conviction scarcely less intense—the conviction that the future will not repeat the past, that time unrolls unidirectionally, that the world is getting on. This second conviction finds quantitative expression in a second principle of thermodynamics. Entropy (Gr. *entrope; en*, in + *trope*, turning) is the state function distinguished by this second principle and, by always increasing in the direction of spontaneous change, the entropy function indicates the "turn" taken by all such change. The foundation of thermodynamics occurred little more than a century ago,[18] precisely when Clausius first brought together the two principles he later stated in the aphorism:

> Die Energie der Welt ist constant.
> Die Entropie der Welt strebt einem Maximum zu.

If a speeding lead bullet is stopped by an unyielding (and thermally insulated) sheet of armor, the gross *kinetic energy* of the bullet is converted into *internal energy* that manifests itself in a rise of temperature. But we never find that equal bits of lead, heated to the same temperature, suddenly cool down and move off with the velocity of bullets—though such a development would be perfectly compatible with the first principle of thermodynamics. Here is a striking but not atypical example of the entirely excessive degree of latitude left open by the first principle, which fails to exclude a great variety of changes never found in practice. As chemists we seek to predict the direction in which a reaction would proceed in reaching equilibrium, but either direction is equally compatible with the first principle. Only with the acquisition of a second principle can we put arrows into our equations before the reactions are tried.

What determines the direction of spontaneous change? Everyday experience suggests certain special cases about which we feel no doubt. Whenever we encounter differences in certain *intensive properties,* we easily predict the direction of spontaneous change as that which eliminates the differences. A difference of *pressure* on the two sides of a piston produces spontaneous movement of the piston in the direction (and to the extent) that yields equalization of pressures. A difference of *temperature* evokes a flow of heat from the hot body to the cold body until at last the temperatures become the same. A difference of *electric potential* sparks a transfer of charge in the direction (and to the extent) required to equalize the potentials. And so on for differences of *concentration* (or, much better, "chemical potential") and the like.

Amply useful though they are, these generalizations are far from being everywhere sufficient. Most obviously, they tell us nothing whatever about the direction of spontaneous chemical reaction, which profoundly changes even systems throughout which there reigns initially complete uniformity of pressure, temperature, electric potential, and concentration(s). One other possibility may then suggest itself. In purely mechanical systems the direction (and extent) of spontaneous change is ordinarily easy to predict—as that which reduces to a minimum the potential energy of the system. At one time an analogous rule was thought capable of predicting the direction of spontaneous chemical change: the spontaneous reaction is that which minimizes the "heat content" of the system. But today we recognize endothermic reactions that occur spontaneously, and so also exothermic reactions that do *not* occur spontaneously. We must then dig a great deal deeper to find a criterion competent to predict the direction (and extent) of spontaneous chemical change. However improbably, we best begin this excavation with an inquiry into the efficiency of heat engines.

CARNOT CYCLE

In a properly operated firebox, a ton of coal yields some large number of kilocalories at some high working temperature. Taking this as heat input, what kind of engine will deliver a maximum output of mechanical work? Consider only the larger questions, entirely apart from constructional details that determine frictional losses, etc. Does an old-fashioned reciprocating engine do as well with steam as it would with ether, or silicone, as working fluid? Does any reciprocating engine do as well as a turbine, and is mercury vapor superior to steam as working fluid in a turbine? Instead of burning the coal in a firebox, would it better be ignited only after spraying it, as a fine dust, into a Diesel-type engine? What about less familiar types of heat engine, e.g., the kind of bar-expansion engine used to rewind the spring in some clocks? Among the myriad possibilities, which is the best in principle? Almost a century and a half ago this question was answered first, and for all time, by Sadi Carnot.

Seeking to analyze the factors that determine the efficiency of actual heat engines, we follow Carnot in conceiving a strictly theoretical engine that is particularly amenable to analysis. One obvious specification for this engine is that it must function *reversibly*, and thus yield the absolute maximum work output of which it is capable in the absence of all secondary losses due to friction, improper harnessing of force to load, etc. A further crucial specification is that (like all actual engines) our theoretical engine must work in a *closed cycle*. At the end of each such cycle the engine itself returns again to its initial state, so that *all* the changes produced by the cyclic function of the engine are localized in the surroundings with which it exchanges heat and work. To determine the efficiency with which the engine converts a heat input into a work output, we have then only to evaluate the changes it produces in its surroundings. This evaluation proves unexpectedly easy if we are intelligent in our choice of the particular closed cycle of changes we subject to analysis. We find nothing better than what is still known as the Carnot cycle: a four-step alternating sequence of reversible adiabatic and isothermal changes, which is just about the simplest way of combining such changes in a closed cycle.

One of the great virtues of the Carnot cycle is its potential applicability to *any* working substance. But we begin by applying it only in the by-now familiar context of an ideal gas undergoing reversible isothermal and adiabatic changes. And at this point we temporarily modify our symbols. To facilitate expression of a series of inequalities, for the rest of this section we will use Q to symbolize $|q|$, the absolute value of the heat term; and W to symbolize $|w|$, the absolute value of the work term. When the system gains heat or does work, we will write $+Q$ or $+W$; when the system loses heat or has work done on it, we will write $-Q$ or $-W$. Thus the *direction* of each transfer of heat and work will be indicated unequivocally by an attached algebraic symbol.

We take as our system a cylinder-piston device charged with ideal gas. By suitably pairing reversible adiabatic and reversible isothermal steps, we can link any given initial state of the system with any given final state. As noted also in Chapter 1, variation in the *order* of the steps yields the two distinct paths exhibited[19] in Fig. 23.

Along path (i) the first reversible expansion is isothermal: doing work against an external pressure only infinitesimally less than the pressure of the gas in the system, the latter absorbs heat $+Q_H$ from a "heat reservoir" (perhaps a huge block of metal) imagined as large enough to retain a steady temperature only infinitesimally greater than that (T_H) of the system. There then follows a reversible adiabatic expansion during which the temperature of the gas drops from T_H to T_L as, without any further

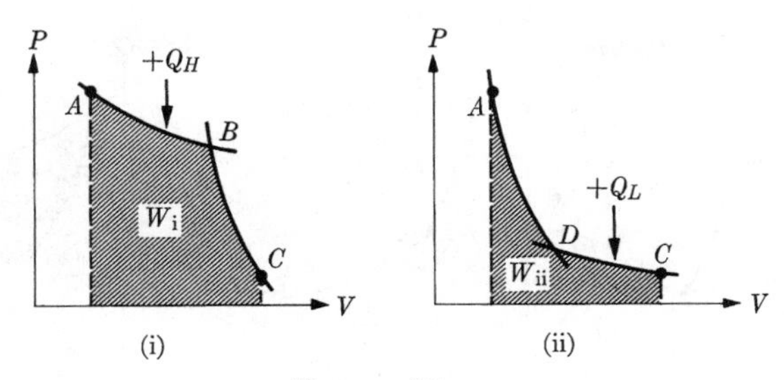

FIGURE 23

absorption of heat, the system delivers some further work output at the expense of its own internal energy.

Along path (ii) the first reversible expansion is adiabatic, and the temperature of the system here drops from T_H to T_L. There then follows a reversible isothermal expansion, during which the system absorbs heat $+Q_L$ from a second heat reservoir supposed large enough to retain a steady temperature only infinitesimally greater than that (T_L) of the system.

Along the two paths, the total work outputs (W_i and W_{ii}) are represented by the shaded areas in the figure, inspection of which leads us to write

$$W_i > W_{ii}.$$

Since both paths connect the same initial state (A) with the same final state (C), the system undergoes exactly the same energy change (ΔE) along either path. As expressed in equation (2), the first principle then entails that

$$Q_H - W_i = \Delta E = Q_L - W_{ii}.$$

Combining this equation with the preceding inequality, we find

$$Q_H > Q_L.$$

And now it is time to think how the four reversible steps displayed in Fig. 23 can be combined in a closed cycle.

Carnot engine and refrigerator

A closed cycle of changes can be achieved by joining path (i) with the inverse of path (ii), which we show as path (ii*) in Fig. 24. Having at the outset stipulated reversible performance of all changes, we know that the *magnitudes* Q_L and W_{ii} must remain the same in (ii*) as in (ii). But the *signs* are of course different: where along (ii) the expansion ADC requires a heat input $+Q_L$ and yields a work output $+W_{ii}$, along (ii*) the compression CDA is brought about by a work input $-W_{ii}$ and yields a heat output

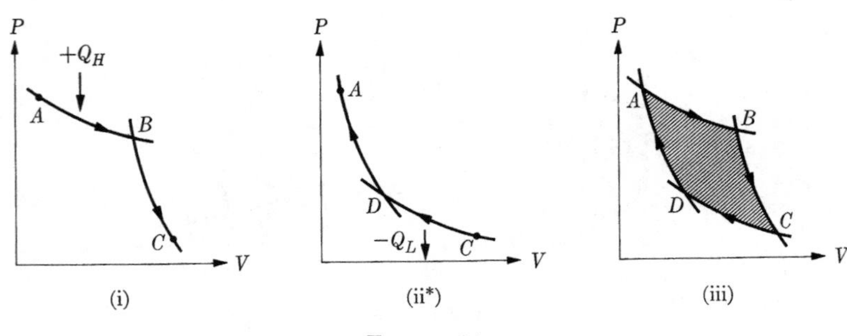

FIGURE 24

$-Q_L$. What then will be the *net* changes produced by the (clockwise) passage of the system once around the cycle formed by joining path (i) with path (ii*)?

By subtracting the shaded area in Fig. 23(ii) from the shaded area in Fig. 23(i), we learn that in Fig. 24(iii) the shaded area that fills the cycle represents its *net work output* ($+W$). The *net heat input* is of course $(Q_H - Q_L)$. For the system itself, passage all around the cycle, back to the starting point at A, yields a net change $\Delta E = 0$. Calling now on the first principle, we substitute in the equation $q - w = \Delta E$ to find

$$(Q_H - Q_L) - W = \Delta E = 0$$

or

$$W = (Q_H - Q_L).$$

With $Q_H > Q_L$, $(Q_H - Q_L)$ is a *positive* quantity here properly associated with what must also be a *positive* term W ($= W_{\text{i}} - W_{\text{ii}}$, where $W_{\text{i}} > W_{\text{ii}}$). Thus the joining of paths (i) and (ii*) constitutes an *engine cycle*, converting a net heat input $(Q_H - Q_L)$ into a net work output $+W$. For the *efficiency* of this cycle the most meaningful expression will be the fraction W/Q_H—since what interests us is the amount of work recoverable per unit input of heat delivered at high temperature by the combustion of some more or less costly fuel. And so we are led to write:

$$\text{Engine efficiency} \equiv \frac{W}{Q_H} = \frac{Q_H - Q_L}{Q_H} = 1 - \frac{Q_L}{Q_H}. \tag{19}$$

To determine the efficiency of the Carnot engine, we have then only to evaluate the ratio Q_L/Q_H. Before attempting this calculation, however, we should note that, by a simple reversal of its cycle, the Carnot engine is convertible to function as a refrigerator.

To construct an identical closed cycle traversed in the opposite (counterclockwise) direction, we have only to join the original path (ii) with the inverse of path (i), which we show as path (i*) in Fig. 25. All steps having

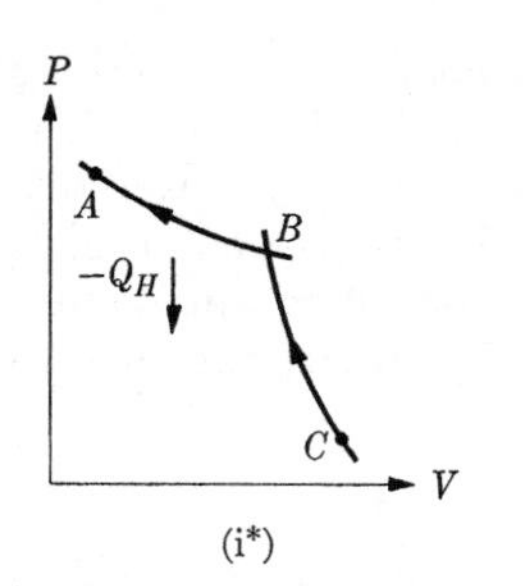

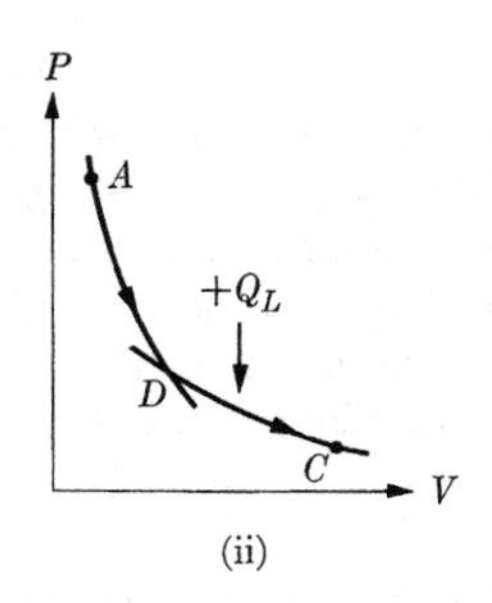

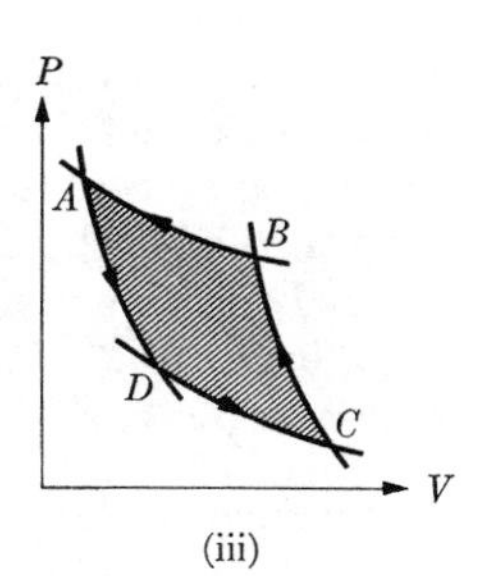

FIGURE 25

been stipulated as reversible, the shift from (i) to (i*) changes only the signs, *not* the magnitudes of the heat and work terms. Subtracting the shaded area in Fig. 23(i) from the shaded area in Fig. 23(ii), we learn that the shaded area in Fig. 25(iii) now represents a *net work input* $(-W)$ where before we had the work output $+W$. And where before we had a net heat input $+(Q_H - Q_L)$, now we have a *net heat output*

$$Q_L - Q_H = -(Q_H - Q_L).$$

For the system itself, on completion of the cycle $\Delta E = 0$. Again substituting in the equation $q - w = \Delta E$, we find

$$-(Q_H - Q_L) - (-W) = \Delta E = 0$$

or

$$-W = -(Q_H - Q_L).$$

To the negative quantity $-(Q_H - Q_L)$ there duly corresponds the work input $-W$ required to drive what we see is now a *refrigerator cycle*. With $\Delta E = 0$ for the refrigerator itself, the magnitude of the work input $(-W)$ is the margin by which the heat $(-Q_H)$ rejected at high temperature *must* exceed in magnitude the heat $(+Q_L)$ taken up at low temperature. This is indeed precisely the message of the last equation, which we can easily combine with the corresponding equation for the heat engine—by writing

$$\pm W = \pm(Q_H - Q_L).$$

All the difference between heat engine and refrigerator is here reduced to a difference of sign. The difference between the (+) work output we expect from the engine and the (—) work input we must make to the refrigerator thus emerges from the difference between the clockwise direction of the engine cycle and the counterclockwise direction of the refrigerator cycle.

Throughout this section the only working substance explicitly under discussion has been the ideal gas. But a suitable combination of adiabatic and isothermal changes should, it appears, yield for any other working substance a closed cycle to which all the previously derived equations would apply. So speculating, we reach the immediate vicinity of Carnot's central insight: the maximum efficiency of an ideal heat engine depends on the boiler temperature (T_H) and the exhaust or condenser temperature (T_L), but is *totally independent* of the mechanism(s) of the engine and the nature of the working substance(s) it may contain.

Carnot theorem

When operated reversibly between given temperatures T_H and T_L, all heat engines have exactly the same limiting efficiency. This theorem we demonstrate by a species of *reductio ad absurdum*. Let us suppose that between the temperatures T_H and T_L there operate two reversible engines (X and X') of which—*contrary* to the theorem—X is more efficient. We will now adjust the size of the engines until, their difference in efficiency notwithstanding, both X and X' deliver the same amount of work *per cycle*. We then couple X and X' as shown in Fig. 26: functioning as an engine, X here drives X' in reverse, as a refrigerator. What will be the result?

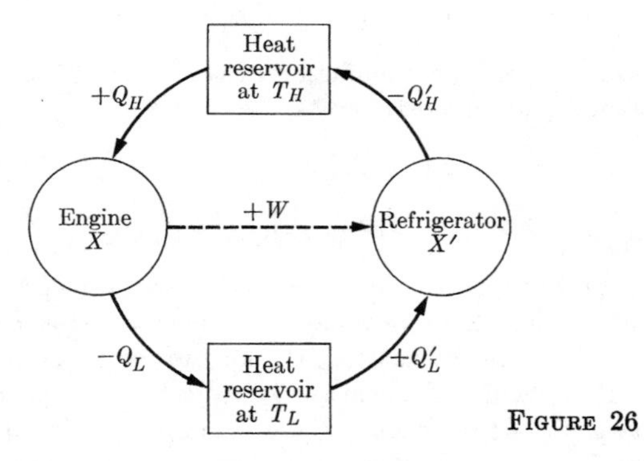

FIGURE 26

In each cycle, X draws heat Q_H at temperature T_H, rejects heat $-Q_L$ at temperature T_L, and delivers a net work output

$$+W = +(Q_H - Q_L).$$

In each cycle X' draws heat Q'_L at temperature T_L, rejects heat $-Q'_H$ at temperature T_H, and demands a net work input

$$-W' = -(Q'_H - Q'_L).$$

Our prior adjustment of the sizes of X and X' assures that the per-cycle work production (W) is perfectly compensated by the per-cycle work consumption ($-W'$), so that

$$W - W' = 0.$$

Hence

$$(Q_H - Q_L) - (Q'_H - Q'_L) = 0$$

and

$$Q_H - Q_L = Q'_H - Q'_L. \tag{a}$$

When acting as an engine, X' has been supposed to have a lower efficiency than X, and equation (19) permits us to express this disparity algebraically, as

$$\frac{Q_H - Q_L}{Q_H} > \frac{Q'_H - Q'_L}{Q'_H}. \tag{b}$$

Comparing equations (a) and (b), we see that

$$Q_H < Q'_H. \tag{c}$$

But then, in view of equation (a), it also follows that

$$Q_L < Q'_L. \tag{d}$$

What is the message of these last two equations? For the system constituted by our coupled engine-refrigerator, operation around the closed cycle leaves $\Delta E = 0$. And zero represents also the net work delivered or consumed in that operation. Thus we have localized in the heat reservoirs *all* the changes produced by operation of the system, and these are indeed truly remarkable changes. From equation (d) we learn that the heat drawn by the refrigerator from the low-temperature (T_L) reservoir *exceeds* the heat delivered to that reservoir by the engine. From equation (c) we learn that the heat delivered by the refrigerator to the high-temperature (T_H) reservoir *exceeds* the heat drawn from that reservoir by the engine. Drawing on no external source of work, the self-sufficient engine-refrigerator system thus encompasses an "uphill" flow of heat, from a cold to a hot body. But such a process is absolutely unknown to us, and is absolutely contrary to an immense body of experience that teaches us to expect the spontaneous flow of heat to be from hot to cold, and never the reverse. Observe too that, by a spontaneous flow of heat from cold to hot, we could transfer heat from the enormous heat reservoir represented by the ocean to another (T_H) reservoir hot enough to power, say, an ordinary steam engine with its condenser at ocean temperature (T_L). Though this arrangement would involve no violation of the first principle of thermodynamics, all the many efforts to build perpetual motion machines of this ("second") kind have eventuated in a failure now so pronounced it teaches us the utter impossibility of all such devices.[20]

We began with the provisional assumption that, of two engines operating reversibly between temperatures T_H and T_L, one is more efficient than the other. In the last paragraph we found that this assumption leads ineluc-

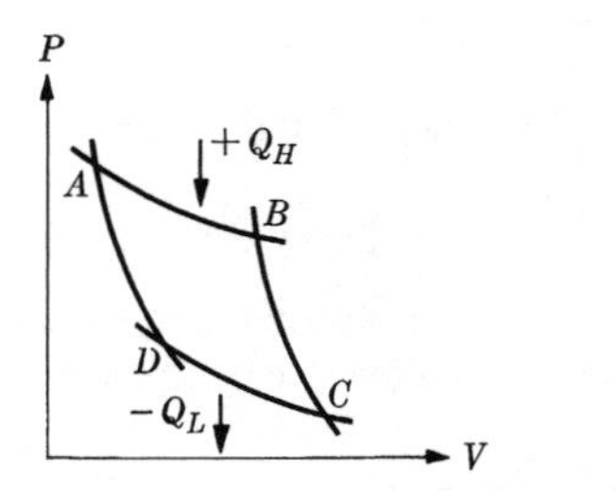

FIGURE 27

tably to conclusions flatly contradicted by a great mass of empirical evidence. We are thus driven to reject the provisional assumption that leads to such untenable conclusions. That is, of the two engines operating reversibly between T_H and T_L, *neither* can be more efficient than the other. And so we have arrived at Carnot's theorem: when operated reversibly between temperatures T_H and T_L, *all* heat engines have the *same* efficiency. The functional dependence of that efficiency on the values of T_H and T_L can then be evaluated for *any* reversible engine we please—and we choose of course to work with an engine containing, of all working substances, that most tractable theoretically: the ideal gas.

The efficiency formula

In Fig. 27 we display the operating cycle of a Carnot engine containing one mole of ideal gas, and delivering a net work output

$$W = Q_H - Q_L.$$

Now Q_H is simply the heat absorbed during the reversible isothermal expansion of the gas from volume V_A to V_B. And $-Q_L$ is the heat rejected during the reversible isothermal compression of the gas from V_C to V_D. For these two changes, equation (18) at once yields:

$$Q_H = RT_H \ln \frac{V_B}{V_A} \qquad \text{and} \qquad -Q_L = RT_L \ln \frac{V_D}{V_C} \cdot$$

Note that, with $V_D < V_C$, the right side of the second equation will be *negative*, as it clearly should be.

Taking the ratio W/Q_H as our definition of engine efficiency, we substitute to find:

$$\text{Engine efficiency} \equiv \frac{W}{Q_H} = \frac{Q_H - Q_L}{Q_H} = \frac{RT_H \ln \dfrac{V_B}{V_A} + RT_L \ln \dfrac{V_D}{V_C}}{RT_H \ln \dfrac{V_B}{V_A}}$$

$$= \frac{T_H \ln \dfrac{V_B}{V_A} - T_L \ln \dfrac{V_C}{V_D}}{T_H \ln \dfrac{V_B}{V_A}}.$$

Now a quite simple derivation* suffices to show that

$$\frac{V_B}{V_A} = \frac{V_C}{V_D},$$

and consequently that

$$\text{Engine efficiency} = \frac{W}{Q_H} = \frac{Q_H - Q_L}{Q_H} = \frac{T_H - T_L}{T_H}. \tag{20}$$

Equation (20) is the consummation of all that has preceded it in this chapter. Before turning to its immense implications, we pause for brief consideration of its direct bearing on the problem that most concerned Carnot: the efficiency of heat engines.

From equation (20) we see that even an ideal heat engine can convert into work output (W) only a fraction of the heat (Q_H) supplied to it at high temperature. A perfect steam engine working between 127° and 27°C will at best yield

$$\text{Efficiency} = \frac{W}{Q_H} = \frac{400 - 300}{400} = 0.25, \qquad \text{or} \qquad 25\%.$$

The challenge of this wretched performance prompts three ideas. Contemplating the efficiency expression

$$\frac{W}{Q_H} = \frac{T_H - T_L}{T_H},$$

our first thought may be to operate heat engines with their exhaust temperatures set as far as possible below the ambient temperature of the surroundings. Problem 27 invites a demonstration that this superficially promising effort must be wholly fruitless. Second: we may think to operate heat engines with input temperatures (T_H) set as high as possible. Shifts

---

* Lying on the same adiabatic, points B and C must also be linked by equation (15):

$$T_H(V_B)^{R/C_V} = T_L(V_C)^{R/C_V} \qquad \text{or} \qquad \frac{T_H}{T_L} = \left[\frac{V_C}{V_B}\right]^{R/C_V}.$$

With points A and D both lying on another adiabatic, equation (15) yields for them:

$$T_H(V_A)^{R/C_V} = T_L(V_D)^{R/C_V} \qquad \text{or} \qquad \frac{T_H}{T_L} = \left[\frac{V_D}{V_A}\right]^{R/C_V},$$

whence it follows at once that

$$\frac{V_C}{V_B} = \frac{V_D}{V_A} \qquad \text{or} \qquad \frac{V_C}{V_D} = \frac{V_B}{V_A}.$$

from steam to other working fluids in part reflect the fruitful effort to push T_H to ever higher values. Third: however high T_H may be pushed, we see that some part of the heat (Q_H) transferred, from burning fuel to heat engine, is predestined for rejection as Q_L. This is the inflexible ordinance of equation (20) which, however, applies only to *thermal* devices. We may then think better to exploit the energy of the fuel by oxidizing it in a galvanic cell rather than underneath the boiler of a heat engine. Much recent research on fuel cells amply demonstrates both the potentially high efficiency and the great practical difficulties associated with the direct recovery of electrical power from oxidizable fuel.[21]

Precisely because heat engines are as inefficient as they are, heat pumps can be enormously advantageous. Consider how we might best "heat with electricity." At first sight we may see no possibility of surpassing the 100%-efficient conversion, of electric energy into heat, so easily achieved by passing electricity through resistance coils. But a quite different way of exploiting the electrical energy is far more advantageous in principle though, because of high capital costs, not always in practice. In principle the electrical energy will best be used to drive a "heat pump," i.e., a refrigerating machine that abstracts heat from the outdoors and delivers it at a higher indoor temperature. To drive the pump we must make the (electrical) *work input* $-W$, and $-Q_H$ will properly symbolize the *heat delivered* by the pump at high temperature. Our last equation, expressing the efficiency of a heat engine, can then easily be recast as an expression of the effectiveness of a heat pump:

$$\frac{-Q_H}{-W} = \frac{T_H}{T_H - T_L}.$$

Suppose, for example, that we seek to maintain an inside temperature of 70°F (= 21°C = 294°K) by pumping heat from an outdoors where the temperature has fallen to 0°F (= −18°C = 255°K). Then:

$$\frac{-Q_H}{-W} = \frac{294}{294 - 255} = \frac{294}{39} \cong \frac{7.5}{1}.$$

Thus for every unit of electric energy, invested now as work used to drive a heat pump, we here recover not just 1 but 7.5 units of energy delivered as heat.[22]

The absolute thermodynamic scale of temperature

To this point we have used the familiar ideal-gas scale of temperatures, based on a particular property (thermal dilation) of a particular group of (gaseous) thermometric substances. This scale is notably arbitrary, in that nothing but human choice excludes myriad other temperature scales, based on the same or different properties of quite different (liquid and solid) thermometric substances. To escape any such element of arbitrariness, however, we need only follow the lead of Lord Kelvin, who first discerned in Carnot's theorem the basis for a truly *absolute* scale of temperature.

On Carnot's theorem, all heat engines working reversibly between two given temperatures must have the *same* efficiency, which relation (19) equates with $W/Q_H = (Q_H - Q_L)/Q_H = 1 - (Q_L/Q_H)$. That is, *regardless* of the identity and state (gaseous, liquid, or solid) of the contained working substance(s), every ideal engine working between the same pair of temperatures must yield the same value for the ratio Q_L/Q_H. Now Q_L and Q_H are in principle empirically determinable quantities of heat rejected to and drawn from cold and hot reservoirs respectively. Thus we have an empirical foundation for the absolute thermodynamic scale of temperature we define by writing

$$\theta_L/\theta_H \equiv Q_L/Q_H,$$

where θ_L and θ_H symbolize the temperatures of the reservoirs on the θ-scale just defined. To be sure the usefulness of this absolute scale may seem hopelessly impaired by the nonexistence of any of the ideal reversible engines that would yield the Q-ratios from which θ-ratios are to be inferred. But from equation (20) we can at once read off the saving message,

$$T_L/T_H = Q_L/Q_H.$$

Comparing the last two equations, we see that θ-values, on the absolute thermodynamic scale of temperature, are directly *proportional* to T-values on an ideal-gas temperature scale that demands no use of ideal heat engines, or even of ideal gases. If then we agree that 273.16° is also to represent the θ-value for the triple point of water, we thereby assure that the absolute scale will everywhere *coincide* with the ideal-gas scale.[23] Beyond a convenience long taken for granted, gas thermometry can now be seen to offer results having a degree of absoluteness we could not before have assumed.

THE CONCEPT OF ENTROPY

The existence of a new function of state—entropy—is by far the most important implication of equation (20), which we now rewrite in the more suggestive form:

$$\frac{Q_H}{T_H} = \frac{Q_L}{T_L}. \tag{21}$$

Consider the familiar fragments of the Carnot cycle shown in panels (i) and (ii) of Fig. 28. In its reversible passage from state A to state C along route (i), the system absorbs heat $+Q_H$ at temperature T_H; in a similar

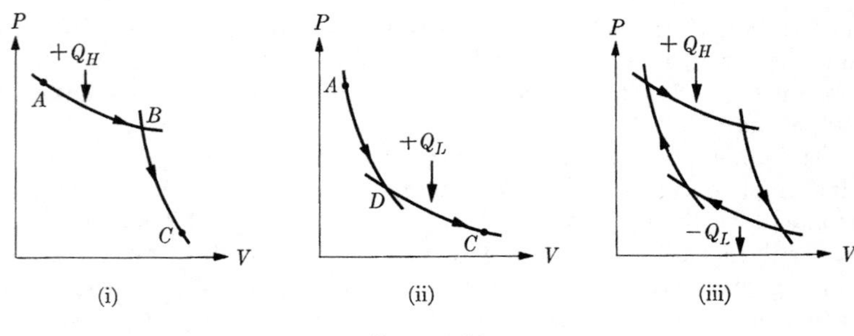

FIGURE 28

passage from A to C along route (ii), the system absorbs a *different* amount of heat ($+Q_L$) at a *different* temperature (T_L). The remarkable assertion of equation (21) is that, the noted differences notwithstanding, the quotient Q/T is the *same* along both routes.

This new finding recalls an earlier discovery that, though both q and w vary from one route to another, for a given change of state the *difference* $(q - w)$ is the same along all routes. That discovery we signalized by creating a new concept, energy, defined by writing $\Delta E = q - w$. Our present finding is that, though both Q and T vary from one route to another, the *quotient* Q/T is the same along both routes shown in the figure. This discovery we signalize by creating a new concept, entropy (S), defined by writing

$$\Delta S = q_{\text{rev}}/T,$$

where the subscript $_{\text{rev}}$ reminds us that all the heat terms under discussion have been, *and must be*, those corresponding to *reversible* performance of the indicated changes.

Is entropy *really* a function of state? Beyond invariance with route, what more do we demand of such a function? When a system resumes its initial condition after passing around a closed cycle of changes, we found that $\Delta E = 0$; and we certainly do demand that any other purported function of state show the same deportment. Will entropy pass this test? As defined above, the entropy of a system undergoing reversible change will increase when the system absorbs heat (for q_{rev} is then positive), and decrease when the system rejects heat (for q_{rev} is then negative). The Carnot cycle of changes shown in Fig. 28(iii) is characterized by the absorption of heat ($+Q_H$) at temperature T_H and the rejection of heat ($-Q_L$) at temperature T_L. And on rearranging equation (21) we find

$$\frac{Q_H}{T_H} + \frac{-Q_L}{T_L} = 0.$$

Thus we learn that, around *this* closed cycle of reversible changes,

$$\Delta S = \sum q_{\text{rev}}/T = 0,$$

which is exactly the behavior to be expected if S symbolizes a genuine function of state.

So far so good, but entropy might still fail to be a *significant* function of state if it applied only to changes that fit the very special Carnot pattern of alternating isothermal and adiabatic steps. Will entropy continue to behave as a proper function of state in reversible changes generally?

Entropy in reversible change

In Fig. 29 the smooth heavy line depicts a closed cycle of reversible changes undergone by *any* substance—gas, liquid, or solid—involved in *any* series of heat exchanges, including exchanges extended over entire continuous *ranges* of temperature. Let the adiabatic and isothermal behavior of the substance be known or ascertainable. We then begin by lacing the figure, as shown, with an array of the substance's *adiabatics*. We continue by drawing in those segments of the substance's *isotherms* that best approximate each short section of the actual cycle. The closer the adiabatic lacing with which we begin, the closer will be the approximation to the original smooth curve, which is indeed the limiting outline attained when the isothermal (and adiabatic) segments become infinitesimal in length. We thus reduce the perfectly general cycle of reversible changes to the composite of a set of infinitesimal Carnot cycles, and the rest is easy.

Consider the series of reversible changes that carries the system from state A *via* B to state C. In each infinitesimal isothermal segment along this route, the system absorbs some particular infinitesimal quantity of

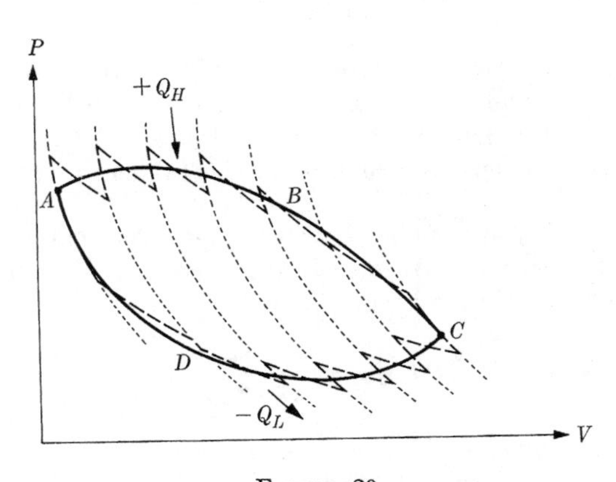

FIGURE 29

heat (Q_H) at some particular temperature (T_H). The paired values of Q_H and T_H presumably differ from each segment to the next. But, taking the quotient of each pair, we can easily sum over all these infinitesimal quotients to find

$$\sum_{ABC} \frac{Q_H}{T_H}.$$

Consider then the series of reversible changes that returns the system from state C *via* D to its initial state A. In each infinitesimal isothermal segment along this route, the system rejects some heat ($-Q_L$) at some temperature (T_L). Summing over all the quotients formed from coordinate values of $-Q_L$ and T_L, we obtain

$$\sum_{CDA} \frac{-Q_L}{T_L}.$$

And now comes the crux of the argument. Though apparently quite independent, the individual quotients in the two series are actually closely related. Each Q_H/T_H quotient along the route ABC is matched with a $-Q_L/T_L$ quotient along the route CDA—in that the two quotients refer to the opposite isotherms of an infinitesimal Carnot cycle. Applying equation (21) to *each* pair of matched quotients, we then find as the sum of *all* such matched quotients,

$$\sum_{ABC} \frac{Q_H}{T_H} + \sum_{CDA} \frac{-Q_L}{T_L} = 0.$$

Thus when the system completes the closed cycle $ABCDA$, its net change of entropy will be

$$\Delta S = \sum_{ABCDA} \frac{q_{\text{rev}}}{T} = 0.$$

Where earlier this result was established only for the special case of a Carnot cycle, we have now demonstrated that $\Delta S = 0$ for a system that has completed *any* closed cycle of reversible changes. In the general case, as in the special, entropy thus continues to behave as a proper function of state.

One obvious corollary of the above demonstration may still be worth remarking. If as before change proceeds from state A *via* B to state C, the entropy change of the system will be

$$\Delta S_{ABC} = \sum_{ABC} \frac{Q_H}{T_H}.$$

If *instead* change proceeds from state A *via* D to state C, with quantities of heat ($+Q_L$) now *absorbed* at temperatures T_L, the entropy change of the system will be

$$\Delta S_{ADC} = \sum_{ADC} \frac{Q_L}{T_L}.$$

But then exactly the same argument of matched quotients will justify the application of equation (21), and lead to the conclusion:

$$\Delta S_{ABC} = \sum_{ABC} \frac{Q_H}{T_H} = \sum_{ADC} \frac{Q_L}{T_L} = \Delta S_{ADC}.$$

Therefore the change of entropy is duly found the same for *all* reversible paths between given initial and final states of the system. An undeniably genuine function of state has then been created when entropy is defined by the equations:

$$\Delta S = \frac{q_{\text{rev}}}{T}, \qquad dS = \frac{q_{\text{rev}}}{T}. \qquad (22)$$

For a finite change at constant temperature — For an infinitesimal change

We can now push one short step further. We found $\Delta S = 0$ for a system that completes a closed cycle of *reversible* changes. But when the system returns to its initial state after traversing a closed cycle of *any* kind of changes, we must still find $\Delta S = 0$, simply because entropy is a function only of the state and not of the history of the system. Again: for a system undergoing some given change of state, we found ΔS the same along all *reversible* paths. But, *whatever* the manner in which the system passes between the given initial and final states, we must still find the same value for ΔS, which symbolizes the change in a function of state. Note however a crucially important difference between the reversible and irreversible changes. When a system changes reversibly, the equation $\Delta S = q_{\text{rev}}/T$ permits a direct determination of ΔS from the amount of heat (q_{rev}) transferred during the change. Though for the system ΔS remains the same when the given change proceeds irreversibly, ΔS is *not* then directly calculable from the heat transfer (q) that accompanies the irreversible change. While at first sight this may appear a little odd, we have already encountered another very similar situation. When a system undergoes a given change of state, ΔE remains the same however the change proceeds; but *only* when the change is brought about at constant volume will the heat transferred (q_V) offer a direct measure of ΔE, by way of the equation $\Delta E = q_V$. Through the first principle, ΔE can still be related indirectly to the heat transfer in a change proceeding with alteration of volume: to what extent can ΔS be related to the transfer of heat that accompanies an irreversible change of state?

Entropy in irreversible change

Consider the general *isothermal* change of a system from state A to state B. Depending on how we conduct it, the change may proceed either reversibly or irreversibly. Let a closed cycle of isothermal changes be constructed by first permitting the system to pass irreversibly from A to B, and then returning it in a reversible manner from B to A. For the irreversible change from A to B we symbolize the heat and work terms as q_{irr}^{AB} and w_{irr}^{AB}, and by q_{rev}^{BA} and w_{rev}^{BA} we symbolize the corresponding terms in the reversible change from B to A. The *net* work transfer $(w_{\text{irr}}^{AB} + w_{\text{rev}}^{BA})$ can easily be related to the *net* heat transfer $(q_{\text{irr}}^{AB} + q_{\text{rev}}^{BA})$, since on completion of the cycle the system itself must show $\Delta E = 0$. Substitution in equation (2) then yields

$$(q_{\text{irr}}^{AB} + q_{\text{rev}}^{BA}) - (w_{\text{irr}}^{AB} + w_{\text{rev}}^{BA}) = \Delta E = 0,$$

$$(w_{\text{irr}}^{AB} + w_{\text{rev}}^{BA}) = (q_{\text{irr}}^{AB} + q_{\text{rev}}^{BA}). \qquad \text{(e)}$$

Can $(w_{\text{irr}}^{AB} + w_{\text{rev}}^{BA})$ be positive? In that event $(q_{\text{irr}}^{AB} + q_{\text{rev}}^{BA})$ must also be positive, and the cyclic *isothermal* process would be characterized by a net absorption of heat and a net output of work. Now the engine-efficiency formula, $(T_H - T_L)/T_H$, indicates that the smaller the difference $(T_H - T_L)$, the smaller will be the net work output recoverable from a given heat input. In the limit $(T_H - T_L) \to 0$ even an ideal reversible engine becomes incapable of delivering a positive net output of work. In our isothermal cyclic process, therefore, $(w_{\text{irr}}^{AB} + w_{\text{rev}}^{BA})$ *cannot* be positive.

Consider the possibility that

$$w_{\text{irr}}^{AB} + w_{\text{rev}}^{BA} = 0.$$

We now recall that actual reversal of a reversible process alters only the *signs* and not the magnitudes of the heat and work terms. Symbolizing by w_{rev}^{AB} the work term in the *reversible* change from state A to state B, we can then write $w_{\text{rev}}^{BA} = -w_{\text{rev}}^{AB}$. Substitution in the last equation then produces

$$w_{\text{irr}}^{AB} - w_{\text{rev}}^{AB} = 0 \qquad \text{or} \qquad w_{\text{irr}}^{AB} = w_{\text{rev}}^{AB}.$$

But such an equality of work terms would reduce the dichotomy reversible–irreversible to a distinction without a difference. The differences being demonstrable, we must reject both the last equation and the assumption from which it derived: $(w_{\text{irr}}^{AB} + w_{\text{rev}}^{BA})$ *cannot* be zero.

Neither positive nor zero, $(w_{\text{irr}}^{AB} + w_{\text{rev}}^{BA})$ can only be negative, so that

$$w_{\text{irr}}^{AB} + w_{\text{rev}}^{BA} < 0,$$

$$w_{\text{irr}}^{AB} - w_{\text{rev}}^{AB} < 0,$$

$$w_{\text{irr}}^{AB} < w_{\text{rev}}^{AB}.$$

For *any* given isothermal change, then,

$$w_{\text{irr}} < w_{\text{rev}}.$$

We have thus found a persuasive general demonstration for a proposition earlier written (on p. 20) only on the strength of its intuitive appeal and a few illustrative examples.*

What most concerns us now is not the relation of the w's but, rather, the relation of the q's. Having established that $(w_{\text{irr}}^{AB} + w_{\text{rev}}^{BA})$ is negative, from equation (e) we now know also that

$$0 > q_{\text{irr}}^{AB} + q_{\text{rev}}^{BA},$$

$$0 > q_{\text{irr}}^{AB} - q_{\text{rev}}^{AB},$$

$$q_{\text{rev}}^{AB} > q_{\text{irr}}^{AB}.$$

Letting T symbolize the temperature throughout the isothermal cycle, we have

$$q_{\text{rev}}^{AB}/T > q_{\text{irr}}^{AB}/T.$$

But, from the definition $\Delta S = q_{\text{rev}}/T$, we conclude that

$$\Delta S^{AB} > q_{\text{irr}}^{AB}/T.$$

And so we arrive at the sought-for relation between ΔS and q when a system undergoes *any* irreversible isothermal change:

$$\Delta S > q_{\text{irr}}/T.$$

Often referred to as the Clausius inequality, this important relation can easily be united with the definition of ΔS, by writing

$$\Delta S \geq q/T. \tag{23}$$

Here the equality refers to reversible change, in which q must represent q_{rev}; the inequality, to irreversible change, in which q represents q_{irr}. This bimodal relation refers explicitly only to isothermal changes, but extension to *non*isothermal change presents no problem. We have only to conceive that change as proceeding through a series of infinitesimal steps in each

---

* As indicated in problem 32, the same line of argument is also competent to demonstrate two important related propositions. First: *no* manner of conducting an isothermal change recovers from it more work than is recoverable when the change proceeds reversibly. Second: when a given change of state can proceed along radically different isothermal routes, the maximum work recoverable along all such routes is exactly the *same*.

of which the temperature is effectively constant. For the ith such step we can then write relation (23) in differential form, as $dS_i \geq q_i/T_i$. Summing over all the infinitesimal steps that together encompass the finite nonisothermal change, we find for the system

$$\Delta S \geq \sum(q_i/T_i).$$

Consider now a system enveloped by rigid, impermeable, *insulating* walls. When only $P\,dV$ work comes in question, this will be a system that can enter into no exchanges of heat or work with the world outside itself. How will the entropy vary when any change takes place in this *isolated system?* Whether we apply the last relation to a change supposed nonisothermal, or instead apply relation (23) to a change assumed isothermal, the result is in either case the same. For an isolated system, with $q = 0$ *by definition*, we find

$$\Delta S \geq 0. \tag{24}$$

Again the equal sign applies to reversible change which, in an isolated system, produces no change of entropy. And the inequality again applies to irreversible change which, in an isolated system, always increases the entropy.

Now when (in the first chapter) we first discussed "reversibility," we found it an ideal limit sometimes approached but never attained by any change actually observed to occur. Hence the last equation teaches us that any change *actually occurring* in an isolated system must increase the entropy thereof. Conceiving the universe as an isolated system, we may then say with Clausius that its entropy "strebt ein Maximum zu." Thus we come to recognize entropy as that directional pointer in search of which we began this chapter. For an isolated system the direction of spontaneous change is always predictable—as the direction in which change produces an increase of entropy. But the great importance of this conclusion may at this point be wholly overshadowed by the loom of three as yet unanswered questions. (1) What *is* entropy? (2) How are we actually to determine the variations in entropy associated with alternative directions of all sorts of changes? (3) How are we to formulate a directional criterion that will apply to chemical reactions, as they normally occur in *non*isolated systems? In the remainder of this chapter and the beginning of the next, we deal in order with these three questions.

What is entropy?

The first principle of thermodynamics poses the concept "energy"; the second principle, the concept "entropy." Feeling that we know what energy *is*, we demand to know what entropy *is*. But now, in point of fact, *do* we really know what energy is? The classical dichotomy is matter *vs.* energy, and energy may then be defined as whatever produces heat. But in the early 20th century this dichotomy was undermined by recognition of the interconvertibility of mass and energy, and to the question "What is energy?" we can now give only the unsatisfactory reply "It is *everything.*" Yet however great may be our uncertainty about the intrinsic nature of energy, the thermodynamic significance of that concept remains wholly unimpaired. For the purposes of classical thermodynamics, we need say no more than that energy (E) is a function of state defined by the equation $dE = q - w$, and this is a statement of the first principle. In exactly the same way, for the purposes of classical thermodynamics, we need say no more than that entropy (S) is a function of state defined by the equation $dS = q_{\text{rev}}/T$, and this is a statement of the second principle. However obscure the nature of entropy may or may not prove to be, the thermodynamic value of this concept remains the same—simply because the significance of entropy lies not in its intrinsic *nature* but rather in its measurable or calculable *behavior;* not in what it *is* but in what it *does.*

Indeed we don't need to know what energy is, but we do find it satisfying and instructive to use the kinetic-molecular theory to *interpret* internal energy in terms of the kinetic and potential energies of atoms and molecules. Neither need we know what entropy is, but we find it satisfying and instructive to use the kinetic-molecular hypothesis to *interpret* entropy in terms of the "randomness" with which atoms and molecules are distributed in space and in energy states. A simple illustration of the subtle concept of randomness is found in the previously cited example of a bullet abruptly stopped by a sheet of armor plate. The bullet's gross kinetic energy disappears, and in its place appears thermal energy that manifests itself in a rise of temperature. *Before* the impact, all the lead atoms comprising the bullet traveled together, as a unit, because all had a single directed component of motion superposed on their uncoordinated thermal motions. *After* the impact, this directed component is randomized: when the bullet's gross motion vanishes, the constituent atoms acquire an increased energy of random thermal motion, which is reflected in the temperature rise. Observe that this molecular picture renders easily intelligible the striking disparity of the following two cases: (i) a moving bullet, when stopped, becomes hotter; and (ii) a stopped bullet, when heated, is *not* thereby set in motion. This otherwise puzzling asymmetry or unidirectionality grows out of a statistical situation amply familiar in everyday experience. Consider for example that a new deck of cards, factory-packed in a regular arrangement of suits and denominations, is soon randomized by shuffling; but we think it highly improbable that, by further shuffling, we will soon return the pack to its original highly-ordered arrangement.

Interpreting increase in entropy as increase in randomness, often we can see *why* where formerly we saw only *that*. Thus on strictly thermodynamic grounds we saw *that* spontaneous change in an isolated system occurs always in the direction of increasing entropy. On our present interpretation we gain some inkling of *why* this is the case: apart from experiences with playing cards, many other familiar situations make us think it "natural" that spontaneous change should always be associated with increase in what Gibbs called "mixed-up-ness." Our own existence, and the growth and proliferation of highly-organized living things generally, in no way breaches the rule that in an *isolated system* change always proceeds with increase in entropy and, therefore, increase also in randomness or disorder. For the earth is *not* an isolated system: energy pours into it from the sun; energy drains away from it into space. The vast input of solar energy powers the gain in organization displayed in the growth of plants as highly-ordered arrays formed from less highly-ordered nutrients like CO_2 and H_2O. And animals achieve their own characteristic organization only by drawing on highly-ordered arrays of atoms presented in foods deriving directly or indirectly from plants. Only a minute upstream eddy in the ongoing torrent of increasing entropy, the occurrence of living organisms is in no way incompatible with our interpretative association of increase in entropy with increase in disorder.[24]

In the next section we continue to interpret increase of entropy in terms of increase in randomness, disorder, or "spread."[25] Facile but slippery, these verbalisms can be supplanted by the rigorous formulations of statistical mechanics, which has proved itself a powerful ally to thermodynamics. But again we emphasize that, whether highly formal or purely verbal, all these interpretations lie *outside* the realm of a classical thermodynamics that owes its peculiarly high reliability to its independence of all hypotheses about the nature of matter, energy, entropy, and so forth.

EVALUATION OF ENTROPY CHANGES

Whatever may be our *interpretation* of the state-function entropy, in thermodynamics *calculation* of entropy change proceeds always from the defining equation $dS = q_{rev}/T$. Sometimes direct empirical determination of q_{rev} is feasible. But direct measurements will of course yield no figure for q_{rev} in processes as irreversible as the impact of the bullet, the solidification of a supercooled liquid, or the expansion of a gas into a vacuum. No matter! We have then only to conceive a sequence of *reversible* steps that link the initial and final states between which the system is carried by the irreversible change. With entropy established as a function of state, the sum of the entropy changes for the several steps must represent the total entropy change (ΔS) for the system undergoing the irreversible change.

Measuring or (more often) calculating q_{rev} for these steps, we can then write $\Delta S = \sum(q_{rev}/T)$. Should ever we find it difficult to conceive the necessary sequence of reversible steps, *still* there is a way out. For in just such a case we are most likely to find it easy to conceive a series of reversible steps that bring about the return of the system from its final to its original state. As regards the system, ΔS in this inverse process differs only in sign from ΔS for the original irreversible process—the entropy change in which can thus be established.

Isothermal expansion of an ideal gas

Let us take as our system the familiar piston-cylinder device containing n moles of ideal gas. What is the change of entropy experienced by this system when the gas expands isothermally and reversibly from some state 1 to a state 2? Equation (18) offers us an expression for q_{rev} in precisely this expansion, so we can write at once:

$$\Delta S = \frac{q_{rev}}{T} = \frac{nRT \ln (V_2/V_1)}{T} = nR \ln \frac{V_2}{V_1} = nR \ln \frac{P_1}{P_2}. \qquad (25)$$

This equation teaches us that, in the reversible isothermal expansion of an ideal gas, increase in entropy ($\Delta S > 0$) goes with increase in volume—since $V_2 > V_1$ means $\ln V_2/V_1 > 0$. Suppose now that the very same isothermal expansion were to occur irreversibly, by expansion of the gas into a vacuum. In *this* expansion $q = w = 0$ but, since entropy is a function of state, ΔS for the system undergoing this expansion is exactly the *same* as that given by equation (25) for the reversible expansion bridging between the same initial and final states. To find the *difference* in the results of the reversible and irreversible processes, we must of course *not* look to the *system*, which is to be left in the same final state by either process. Instead we must look to the *surroundings* which, after exchanging with the system necessarily different amounts of both heat and work, will be left in quite *different* states at the conclusion of the reversible and irreversible processes.

▶ *Example 1*

What is the change in the entropy of one mole of ideal gas when its volume is doubled by an isothermal expansion that proceeds (a) reversibly or (b) irreversibly into a vacuum?

Solution. (a) For the reversible expansion equation (25) gives

$$\Delta S = nR \ln (V_2/V_1) = (2.30)(1)(1.99) \log 2 = +1.38 \text{ cal/mole-°K}.$$

(b) For the irreversible expansion to the *same* final state, the entropy change of the gas must again be +1.38 cal/mole-°K. ◀

▶ *Example 2*

What is the change in entropy of the surroundings when the volume of one mole of ideal gas is doubled by an isothermal expansion that proceeds (a) reversibly or (b) irreversibly?

Solution. (a) In any reversible isothermal change the surroundings must stand at the same temperature (T) as the system, so by definition:

$$\Delta S_{\text{surr}} = (q_{\text{rev}})_{\text{surr}}/T.$$

Heat lost by the system and heat gained by the surroundings are, though opposed in sign, necessarily the same in magnitude. Hence

$$\Delta S_{\text{surr}} = -(q_{\text{rev}})_{\text{syst}}/T.$$

But, again by definition, $(q_{\text{rev}})_{\text{syst}} = T\,\Delta S_{\text{syst}}$, which means

$$\Delta S_{\text{surr}} = -T\,\Delta S_{\text{syst}}/T = -\Delta S_{\text{syst}}.$$

This is a usefully general equation but, unlike the very similar relation $\Delta E_{\text{syst}} = \Delta E_{\text{surr}}$, the relation of the entropy differences applies only to *reversible* changes. In the present case this equation yields as answer:

$$\Delta S_{\text{surr}} = -1.38 \text{ cal/mole-°K}.$$

(b) Where in the reversible isothermal expansion the uptake of heat by the system leaves the surroundings in a state of lower entropy, in the irreversible expansion into a vacuum $q = w = 0$. At the close of *this* expansion the surroundings will thus stand in the very same state as at the beginning, which means

$$\Delta S_{\text{surr}} = 0.$$ ◀

Examine now a case in which we need no longer consider the surroundings. Imagine an *isolated* system consisting of two mechanically rigid and thermally insulated bulbs, equal in volume and joined by a stopcock. Let us begin with the first bulb evacuated, and the second containing one mole of ideal gas. What will happen when the stopcock is opened? Considered in the light of equation (24), the entropy increase we have associated with such an expansion determines *that* in this isolated system the gas should expand spontaneously. *Why* this should be a spontaneous change we may hope to learn from our (nonthermodynamic) interpretation of the entropy increase, in terms of the greater randomness or disorder of the gas when it is "spread" throughout both bulbs. To recognize as such the greater randomness of the spread system, we need only consider the likelihood that, at some future time, we will again find all the gas reassembled in the original volume V_o, while the other (equal) volume again becomes void. Once the stopcock has been opened, the probability that some one given molecule will be found in V_o is clearly $\frac{1}{2}$—in the same sense that $\frac{1}{2}$ is the probability that a tossed coin will fall heads. The probability that a given second molecule will be found in V_o is also $\frac{1}{2}$, and the probability that *both* molecules will simultaneously be present in V_o is $(\frac{1}{2})^2 = \frac{1}{4}$—even as $\frac{1}{4}$ is the probability that a twice-tossed coin will twice fall heads. For three of either molecules or coins, the probability is $(\frac{1}{2})^3 = \frac{1}{8}$, and so on. When each of N different objects can fall in either of two categories, the total number of possible arrangements is 2^N, and $1/2^N = (\frac{1}{2})^N$ is the probability of that unique result in which *all* happen to fall in the same category.

Consider now that a mole of ideal gas contains 6×10^{23} independent molecules. When, by first turning the stopcock, we connect the gas-charged bulb with the evacuated bulb, we have indeed good reason to expect that the gas will spread itself throughout both bulbs. Among all the possible distributions we have made accessible to the gas, the probability of that one in which all the molecules remain (or come again to re-collect themselves) in the original bulb has the almost inconceivably minute value of $(\frac{1}{2})^{6 \times 10^{23}}$. Though stupendous as an *improbability*, this still differs from a flat *impossibility*, and so affords us some glimpse of what Gibbs was the first to recognize: the essentially statistical character of the second principle of thermodynamics. But, for any macroscopic sample of gas, the balance of probability is so overwhelmingly one-sided that we have not a moment's doubt in predicting the spread of the gas throughout the isolated system. And actually gases *do* expand spontaneously to fill their containers, and *do not* spontaneously recompress themselves.

Gases can of course be recompressed, in *non*isolated systems. Let us then turn back to the original cylinder-piston system, and let us at the same time return from interpretative speculations to strictly thermodynamic considerations. To recompress the gas, we must do work upon it. More than that, in an *isothermal* compression the ideal gas must be permitted to transfer to the surroundings an amount of heat equivalent to the work invested in bringing about the compression. Indeed, from the relation $\Delta S \geq q/T$ we learn that reduction of the entropy of *any* system must always entail a discharge of heat by the system (i.e., $\Delta S < 0$ entails $q < 0$). In reversible or irreversible isothermal compressions to the same final state, ΔS for the system must in either case remain the same. But, by demanding a greater work input from the surroundings and making a greater heat output to the surroundings, the irreversible compression leaves the surroundings in a state different from that in which they are left by the reversible compression.

▶ *Example 3*

When the volume of one mole of ideal gas is halved by a reversible isothermal compression, what is the change of entropy of (a) the gas and (b) the surroundings?

Solution. (a) For the gas that constitutes our system, equation (25) yields

$$\Delta S = nR \ln (V_2/V_1) = (2.30)(1)(1.99) \log (\tfrac{1}{2}) = -4.58 \log 2$$
$$= -1.38 \text{ cal/mole-°K.}$$

(b) The decrease in the system's entropy is necessarily accompanied by a rejection of heat that increases the entropy of the surroundings. For this *reversible* change we have, as in Example 2,

$$\Delta S_{surr} = -\Delta S_{syst} = +1.38 \text{ cal/mole-°K.}$$ ◀

One can much improve his understanding of the concept "reversibility" by examining the general bearing of the solutions to Examples 1 through 3. Consider for example the general message that finds specific expression in the accompanying balance sheet for two closed cycles built on the changes characterized in those Examples.

TABLE 2

	System ΔS_{syst} +	Surround ΔS_{surr} =	System + Surround ΔS_{total}
Rev. expansion	+1.38	−1.38	0.0
+			
Rev. compression	−1.38	+1.38	0.0
=			
Complete cycle I	0.0	0.0	0.0
Irrev. expansion	+1.38	0.0	+1.38
+			
Rev. compression	−1.38	+1.38	0.0
=			
Complete cycle II	0.0	+1.38	+1.38

Any closed cycle will, by definition, return the system to its original state. But only when the entire closed cycle is conducted reversibly will *both* the system and its surroundings be restored at the end to identically the same condition as at the outset. A reversible change, once reversed, leaves *no* trace on the universe. After any irreversible change, on the other hand, even the most ingenious efforts to restore the original condition inevitably leave some residual alteration in either the system or its surroundings.

The irreversibility *creates* an entropy increase that can thereafter never be destroyed but only transferred. To see the same point from a new angle, consider the limited applicability of the equation $\Delta S_{surr} = -\Delta S_{syst}$ in the perspective of the figures given in the last column of Table 2. These figures refer to the total entropy change of the larger ensemble constituted, in imagination, by uniting the primary system with all of the surroundings with which it interacts to any significant extent. All relevant interactions being thus confined *within* the ensemble, it becomes in effect an *isolated* ensemble to which we can apply equation (24), $\Delta S \geq 0$. In reversible change, where the equal sign applies,

$$\Delta S_{surr} + \Delta S_{syst} = \Delta S_{ensemble} = 0,$$

which entails the relation $\Delta S_{surr} = -\Delta S_{syst}$. In irreversible change, where the inequality applies, the perfect compensation expressed by the indicated relation can no longer obtain—as is indeed evident from the figures in our tabulation.

Phase changes

At the normal melting temperature (T_M) of a crystalline solid under a constant pressure of 1 atmosphere, one mole of solid is liquefied by a heat input that represents the molar heat of fusion (ΔH_{fus}). The melting process may be regarded as effectively reversible: if the surroundings fall even infinitesimally below T_M, the liquid begins to solidify—and heat is no longer absorbed but, rather, rejected by the system. For the molar entropy change (ΔS_M) associated with melting, we have then

$$\Delta S_M = q_{rev}/T_M = \Delta H_{fus}/T_M.$$

Figures for ΔS_M are usually of the order of 2 to 4 cal/mole-°K, but occasionally much larger.[26] To rationalize these increases in entropy we need only observe that, in melting, the highly-ordered lattice structure of the solid is broken down to a much less regular arrangement of atoms or molecules in the liquid phase.

▶ *Example 4*

Under 1 atm pressure at 279°K, the reversible crystallization of benzene from its melt proceeds with an entropy change of −8.53 cal/mole-°K. What is the molar heat of fusion of benzene at its melting point?

Solution. Since the process is reversible, the freezing point of the liquid benzene will represent also the melting point of solid benzene. And if ΔH_{fus} symbolizes the molar heat of fusion of benzene, $-\Delta H_{fus}$ will represent the heat liberated during the crystallization of a mole of benzene.

Hence:

$$\frac{-\Delta H_{\text{fus}}}{279} = -8.53 \quad \text{or} \quad \Delta H_{\text{fus}} = 279(853) = 2.38 \times 10^3 \text{ cal/mole.}$$ ◀

At the normal boiling temperature (T_B) of a liquid under a constant pressure of 1 atm, a heat input representing the molar heat of vaporization (ΔH_{vap}) yields an effectively reversible production of one mole of vapor. Symbolizing by ΔS_B the molar entropy change in the vaporization, we write

$$\Delta S_B = \Delta H_{\text{vap}}/T_B.$$

Compared to ΔS_M, a very much larger increase of entropy may be expected to accompany the conversion of a small volume of (still appreciably structured) liquid into a highly dispersed vapor phase in which the distribution of atoms or molecules is completely random. Actually ΔS_B *is* very large and—even more interesting—Trouton's rule calls attention to the fact that, for a great variety of substances, the quotient $\Delta H_{\text{vap}}/T_B$ lies in the vicinity of 21 cal/mole-°K.

▶ *Example 5*

Under a constant pressure of 1 atm at its normal boiling point of 353°K, one mole of benzene is vaporized by a heat input of 7.35 kcal/mole. Does benzene conform to Trouton's rule?

Solution. The entropy of vaporization is here

$$\Delta S_B = 7350/353 = 20.8 \text{ cal/mole-°K}.$$

Evidently benzene conforms quite well to Trouton's rule. ◀

▶ *Example 6*

Does Trouton's rule reflect anything more than the increased spatial spread of particles passing from a compact liquid into a diffuse gas? Strictly speaking this is *not* a thermodynamic question. But we can easily make a crude estimate by calculating ΔS for the expansion of one mole of ideal gas from the initial volume of the liquid phase to the final volume of the vapor phase. In the case of benzene, at $T_B = 353$°K a mole of liquid occupying 90 ml passes over into vapor occupying approximately 29 liters ($= \frac{353}{273} \times 22.4$ liters). Estimate the contribution of spatial spread to the total entropy of vaporization.

Solution. In the expansion of a mole of ideal gas from 90 ml to 29 liters, equation (25) gives as the entropy change

$$\begin{aligned}\Delta S &= nR \ln (V_2/V_1) = (2.30)(1)(1.99) \log 29000/90 = 4.58 \log 320 \\ &= 11.5 \text{ cal/mole-°K}.\end{aligned}$$

Our primitive interpretation in terms of spatial spread thus accounts for half, but *only* for half, of the observed effect. ◀

Trouton's empirical rule falls far short of perfection. Varying from ca. 18 for O_2 boiling at 90°K to ca. 23 for Zn boiling at 1180°K, the quotient $\Delta H_{\text{vap}}/T_B$ only crudely approximates a steady 21 cal/mole-°K. However we can easily rationalize the shortcomings of Trouton's rule, and at the same time improve upon it. We need only exploit our finding that spatial spread is a major element in the total entropy of vaporization. For consider that, though we work always at a uniform pressure of 1 atmosphere, at different boiling temperatures one mole of vapor will be spread over very different volumes. A mole of oxygen, vaporized at its normal boiling point of 90°K, is spread over only 7.4 liters; but a mole of zinc, at *its* normal boiling point of 1180°K, is spread over a volume more than ten times as great. Thus, apart from entropies of vaporization that possibly are the same, the Trouton "constant" involves also entropies of expansion that are most certainly different.[27]

To improve upon Trouton's rule, we should seek to evaluate entropies of vaporization under conditions that assure a more nearly constant "spread" of the gas phase. On Hildebrand's rule, we may operate with whatever is the heat of vaporization (ΔH_{hv}) at the temperature (T_{h}) at which the given liquid yields saturated vapor with a concentration of 1 mole per 22.4 liters ($\cong$ 0.045 mole/liter). One then finds:

$$\Delta H_{\text{hv}}/T_{\text{h}} \cong 20.3 \text{ cal/mole-°K}.$$

The approximation of *this* quotient to the indicated constant is usually excellent: indeed, the figure 20.3 is rigorously applicable both to O_2 and to Zn at a temperature more than 1000°K higher. A few apparent exceptions still exist, e.g., for water and ammonia the "constant" is close to 25. But this is very much what we should expect once we know something about the nature of water and ammonia. Due to "hydrogen bonding," these substances are abnormally associated and structured even in the liquid phase. For such liquids, vaporization involves a greater than average loss of order, and is therefore associated with a greater than average increase in entropy.

Apart from fusion and vaporization, other possible phase changes include transitions between allotropic species like diamond and graphite. But all these changes can be treated in the same way. At a pressure and temperature (T) at which the two phases stand in equilibrium with each other, a constant-pressure process involving some particular heat transfer (ΔH) can be regarded as reversible. For the change in entropy of the system undergoing this process, we write

$$\Delta S = \Delta H/T. \tag{26}$$

Change of temperature

What is the change of entropy when one mole of substance, with heat capacity C_P, is heated at constant pressure from temperature T_1 to temperature T_2? In imagination we resolve this nonisothermal process into a series of infinitesimal steps in each of which the temperature can be treated as effectively constant. Summing over the entire series, we write

$$\Delta S = \sum(q_{\text{rev}}/T) = \int_{T_1}^{T_2} \frac{C_P\,dT}{T}.$$

When C_P varies appreciably over the temperature range concerned, we will have to express C_P as $f(T)$, and integrate the resultant function. But for our purposes it will generally suffice to treat C_P as a constant, and then

$$\Delta S = \int_{T_1}^{T_2} \frac{C_P\,dT}{T} \underset{\text{IF } C_P \text{ is constant}}{=} C_P \int_{T_1}^{T_2} \frac{dT}{T} = C_P \ln \frac{T_2}{T_1}. \quad (27)$$

After substitution of C_V for C_P, the same equation can be used to determine the entropy change accompanying temperature alterations made at constant volume.

▶ *Example 7*

Over a very broad range of temperatures, 7.0 cal/mole-°K can be taken as C_P for carbon monoxide. What is the entropy change when a mole of CO is (a) heated from 100°K to 200°K, or (b) cooled from 1500°K to 750°K?

Solution

(a) $\Delta S = (2.30)(7.0) \log \frac{200}{100} = 16.1 \log 2 = 4.8$ cal/mole-°K.

(b) $\Delta S = (2.30)(7.0) \log \frac{750}{1500} = -16.1 \log 2 = -4.8$ cal/mole-°K.

Observe that, as in the isothermal expansion of an ideal gas, what counts here is not absolute values but only the *ratio* of initial and final values. ◀

In terms of what randomization or spread shall we interpret the increase of entropy with rise of temperature? If the substance expands when heated, some increase of entropy may be associated with the enlargement of the spatial spread. Substantial in the case of gases, this effect is far too small to be a major contributor to the entropy increases accompanying the heating of solids and liquids; and increase in spatial spread can contribute *nothing* when the heating is conducted under (or calculated for) constant-volume conditions. Yet we can easily rationalize the rise of entropy with temperature—in terms of a spreading effect more subtle than the purely spatial spread earlier noted.

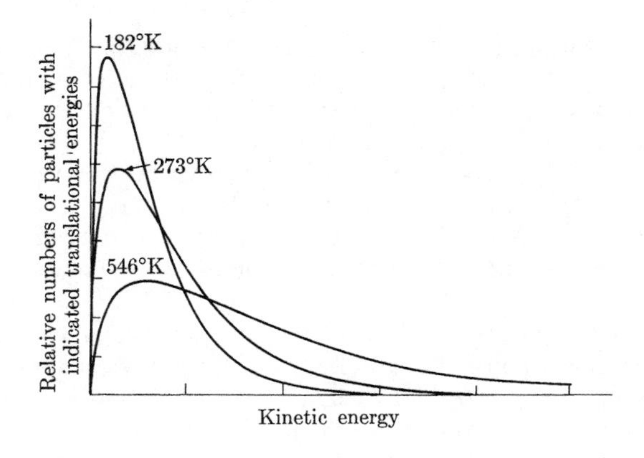

FIGURE 30

In a substance heated at constant volume, and with no phase change, the atoms or molecules "spread" by distributing themselves over a wider range of energetic states. A vivid illustration of this effect is provided in Fig. 30, which shows the distribution of translational energies in an ideal gas at three different temperatures. At low temperature comparatively large fractions of the molecules present are found in a comparatively small range of accessible energy states. As the temperature rises, however, more and more higher energy states become accessible to more and more molecules—which so spread themselves that only a comparatively minute fraction of those present will be found in any small range of energies. A substantial increase in entropy is assignable to this spreading of translational energies. And we can *fully* account for the increase of entropy with

TABLE 3*

Number of the vibrational level	Fraction of the molecules of gaseous NO in each vibrational level as a function of temperature in °K				
	300	600	1000	2000	5000
0	0.9999	0.9886	0.9311	0.7344	0.4020
1	0.0001	0.0113	0.0640	0.1936	0.2373
2	0.0000	0.0001	0.0046	0.0521	0.1413
3		0.0000	0.0003	0.0143	0.0848
4			0.0000	0.0040	0.0513
>4				0.0016	0.0833

* Data of Johnston and Chapman reproduced, by permission, from *J. Am. Chem. Soc.* **55,** 153 (1933).

temperature when, in addition to the spread of translational energies, we allow also for the increased spread of energies associated with rotational, vibrational, and other degrees of freedom. An example of the progressive spread in vibrational energies is shown in Table 3. For a solid at low temperature the rise of entropy with increasing temperature is *entirely* a matter of the spread into higher vibrational energy levels. On this spread Einstein founded an early theory of low-temperature heat capacities that was one of the first important successes of Planck's quantum hypothesis.

The joint occurrence of different kinds of spread is prettily illustrated in the adiabatic expansion of an ideal gas. When the expansion proceeds reversibly, $q_{rev} = q = 0$, and then $\Delta S = 0$. The entropy remains unchanged because an entropy *increment*, due to increased spread of the molecules in space, is here perfectly compensated by an entropy *decrement* due to the decreased spread of molecular energies at the lower temperature assumed by the gas as a result of its expansion. We can see exactly how this compensation comes about if we imagine the adiabatic expansion resolved into the two steps shown in Fig. 21 (on p. 51). When in the first step the gas is cooled at constant volume (V_1), the consequent entropy decrement is given by equation (27) as $C_V \ln (T_2/T_1)$. In the second step the isothermal expansion of the gas produces an entropy increment given by equation (25) as $R \ln (V_2/V_1)$. But rearrangement of equation (d) on p. 52 gives for this reversible adiabatic expansion:

$$C_V \ln \frac{T_2}{T_1} + R \ln \frac{V_2}{V_1} = 0.$$

The perfect compensation, of entropy increase by entropy decrease, no longer obtains if the adiabatic expansion of the ideal gas is to any degree irreversible. Let the gas be expanded as before from volume V_1 to V_2. With the same increase in spatial spread will be associated the same entropy increment as in the reversible expansion. But in the irreversible expansion the gas does less work, loses less internal energy, and so arrives at the end of the expansion with a temperature greater than the above T_2. (This difference in the final states is precisely what we found in the solutions to Examples 7 and 8 on pp. 52–53.) With the smaller temperature drop in the irreversible expansion will be associated a smaller decrement in the spread of molecular energies, and with that in turn an entropy reduction insufficient to counterbalance the entropy increment due to spatial spread. To be sure, though adiabatic, the irreversible expansion *must* show an increase in entropy. For it is now $q_{irr} = q = 0$ and, with $q_{rev} > q_{irr}$, it follows necessarily that $\Delta S > 0$. And on that note we close this series of three sections in which we have everywhere interpreted and *discussed* entropy changes in kinetic-molecular terms, but have everywhere *determined* entropy changes in terms of q_{rev}/T.

Standard entropies and the Nernst heat theorem

We have so far calculated only *differences* of entropy, and only differences of entropy figure in our statement and use of the second principle. But, just as in the earlier case of enthalpies, for convenience we seek to devise a scale of standard entropies that will permit us to assign an individual entropy value to any particular state of any given system. As in dealing with enthalpies, here also we proceed by defining a group of reference states for which entropies are established by convention.

> *Definition:* A value of zero is to be assigned as the entropy of each pure element in a state of complete internal equilibrium at a temperature $T \to 0°K$.

States of internal equilibrium are the implicit presupposition of most thermodynamic statements. That presupposition is here made explicit in order to exclude certain aberrant instances that arise from the extreme slowness with which equilibrium is attained at $T \to 0°K$.

Our entropy scale is now well fixed. But actual determination of entropies on this scale can be enormously simplified because the above definition is powerfully complemented by a further statement:

> *Empirical proposition:* In any isothermal change involving only pure phases in complete internal equilibrium, $\Delta S \to 0$ in the limit $T \to 0°K$.

This is the Nernst heat theorem, sometimes restated in more extreme form as "the third law of thermodynamics."* Complete internal equilibrium is specified for the same reason as above. And, by specifying pure phases, we again exclude solutions which would manifest finite entropies of mixing.

Unlike the definition used to set up a scale of standard enthalpies, the definition on which we have founded a scale of standard entropies does *not* specify zero as the entropy of only the *most stable* form of the element at $T \to 0$. No such specification is necessary because the Nernst heat theorem indicates that, in the limit $T \to 0$, the transition from one to another allotropic form of the same element must yield $\Delta S = 0$. When by definition any one such form is assigned zero entropy, by the Nernst theorem *all* must have zero entropy. And from the same theorem we can draw a far more sweeping inference of very similar form. In the limit

---

* Important in statistical mechanics, the "third law" discharges in thermodynamics no larger role than the Nernst heat theorem—which serves only to establish zero as the value of a constant of integration. Like the two major principles of thermodynamics, the heat theorem is derivable from an empirical record of frustration. That is, where they are underwritten by repeated failures to build operable perpetual motion machines, the heat theorem can be shown to follow necessarily from our repeated experience and firm conviction that, though indefinitely approachable, a temperature of 0°K is unattainable.[28]

$T \to 0$, the formation of a pure compound from its pure component elements must proceed with $\Delta S = 0$. When by definition the elements have been assigned zero entropy, by the Nernst theorem their pure compounds must also have zero entropy in the limit $T \to 0$.

Conclusion: When in states of complete internal equilibrium, all pure elements and all pure compounds have entropy of zero in the limit $T \to 0$.

This conclusion opens the way to a straightforward assignment of standard entropies to all elements and compounds.

Let S^0_{298} and S^0_0 symbolize the molar entropies of a pure substance in its standard states at 298° and 0°K, respectively. In view of the above conclusion, we can express the entropy difference between these states as

$$\Delta S = S^0_{298} - S^0_0 = S^0_{298}.$$

To determine S^0_{298} we have then only to call upon our basic definition of entropy,

$$S^0_{298} = \Delta S = \sum(q_{\text{rev}}/T),$$

where the summation must be extended over all the heat inputs required to bring the substance in question from 0° to 298°K.

In the simplest case of a crystalline solid that preserves the same structure at 298° as in the limit $T \to 0$°K, we write

$$S^0_{298} = \int_0^{298} \frac{C_P \, dT}{T}.$$

Extrapolating toward 0°K values of C_P that fall off very rapidly toward zero, in the region of extreme low temperatures we can proceed either analytically[29] or by graphical integration. At higher temperatures C_P can usually be expressed as an analytical function of T, and there is then no problem to carrying through the indicated integration.

What if, instead of a solid, we have to deal with a substance that melts at some temperature (T_M) less than 298°K? To determine S^0_{298} for the resultant liquid, we need only introduce two additional terms into our calculation of $\sum(q_{\text{rev}}/T)$.

$$S^0_{298} = \int_0^{T_M} \frac{(C_P)_s \, dT}{T} + \frac{\Delta H_{\text{fus}}}{T_M} + \int_{T_M}^{298} \frac{(C_P)_l \, dT}{T}.$$

Here $(C_P)_s$ and $(C_P)_l$ respectively symbolize the heat capacities of the solid and liquid, and the enthalpy of fusion (ΔH_{fus}) is handled as required by equation (26). If the substance boils at some temperature (T_B) less than 298°K, an obvious extension of the same line of argument yields

$$S^0_{298} = \int_0^{T_M} \frac{(C_P)_s\,dT}{T} + \frac{\Delta H_{\text{fus}}}{T_M} + \int_{T_M}^{T_B} \frac{(C_P)_l\,dT}{T} + \frac{\Delta H_{\text{vap}}}{T_B} + \int_{T_B}^{298} \frac{(C_P)_g\,dT}{T}. \tag{28}$$

Given the requisite data on heat capacities,[30] and on heats of fusion, vaporization, etc., we can then calculate S^0_{298} for all elements and compounds. From these, in turn, we can calculate the entropy change (ΔS^0_{298}) in any chemical reaction involving only pure phases, and proceeding under standard conditions at 298°K.[31] For the reaction

$$aA + bB = cC + dD,$$

we will have simply

$$\Delta S^0_{298} = [c(S^0_{298})_C + d(S^0_{298})_D] - [a(S^0_{298})_A + b(S^0_{298})_B].$$

Suppose that, having so determined ΔS^0_{298}, we wished to determine ΔS^0_T for some other temperature. If we simplify only to the extent of assuming no phase changes between 298°K and temperature T, a derivation entirely analogous to that which yielded (Kirchhoff's) equation (11) will yield here

$$\Delta S^0_T = \Delta S^0_{298} + \int_{298}^{T} \frac{\Delta C_P\,dT}{T}.$$

In our earlier discussion of the variation of ΔH^0 with temperature, we found that ΔC_P is ordinarily a rather small quantity. Thus, like ΔH^0, a value for ΔS^0 is usually not much changed by small variations of temperature.

▶ *Example 8*

When by definition we assign zero entropy to the elements at 0°K, we find S^0_{298} for Na(s), Cl_2(g), and NaCl(s) to be respectively 12.2, 53.2, and 17.3 cal/mole-°K. (a) Calculate ΔS^0_{298} for the reaction:

$$\text{Na(s)} + \tfrac{1}{2}\,\text{Cl}_2\text{(g)} = \text{NaCl(s)}.$$

(b) If, on an alternate convention, we were to assign zero entropy to the elements in their standard states at 298°K, what would we then find as the entropy change in the indicated reaction? (c) What conclusion follows from the comparison of the answers to parts (a) and (b)?

Solution. (a) $\Delta S^0_{298} = 17.3 - 12.2 - \frac{1}{2}(53.2) = -21.5$ cal/mole-°K. (b) Symbolizing by $\mathcal{S}$ entropy values based on the alternate convention, for the two elements we write at once $(\mathcal{S}^0_{298})_{Na} = 0$ and $(\mathcal{S}^0_{298})_{Cl_2} = 0$. To determine $\Delta\mathcal{S}^0_{298}$ for the reaction in question, we have then only to find a value for $(\mathcal{S}^0_{298})_{NaCl}$. We do so by calling on the Nernst theorem, which establishes a *difference* of entropy we can formulate with equal propriety as either $\Delta S^0_0 = 0$ or $\Delta\mathcal{S}^0_0 = 0$. On the normal convention, setting $S^0_0 = 0$ for the elements, the result $S^0_{298} = 12.2$ for Na(s) actually means

$$S^0_{298} - S^0_0 = +12.2.$$

This *difference* of entropy must also remain the same wherever we set our zero of entropy. On the alternate convention we can then write

$$(\mathcal{S}^0_{298})_{Na} - (\mathcal{S}^0_0)_{Na} = 0 - (\mathcal{S}^0_0)_{Na} = +12.2,$$
$$(\mathcal{S}^0_0)_{Na} = -12.2 \text{ cal/mole-°K}.$$

And the same line of argument yields

$$(\mathcal{S}^0_0)_{Cl_2} = -53.2 \text{ cal/mole-°K}.$$

For the reaction $Na(s) + \frac{1}{2}Cl_2(g) = NaCl(s)$, as it would hypothetically occur at $T \to 0$°K, the Nernst theorem yields $\Delta\mathcal{S}^0_0 = 0$, so that:

$$\Delta\mathcal{S}^0_0 = (\mathcal{S}^0_0)_{NaCl} - [(\mathcal{S}^0_0)_{Na} + \tfrac{1}{2}(\mathcal{S}^0_0)_{Cl_2}] = 0,$$
$$(\mathcal{S}^0_0)_{NaCl} = +[(-12.2) + \tfrac{1}{2}(-53.2)] = -38.8 \text{ cal/mole-°K}.$$

Now a given datum tells us that for NaCl(s) $S^0_{298} - S^0_0 = +17.3$, and this *difference* of entropy can also be expressed as

$$(\mathcal{S}^0_{298})_{NaCl} - (\mathcal{S}^0_0)_{NaCl} = +17.3,$$
$$(\mathcal{S}^0_{298})_{NaCl} = -38.8 + 17.3 = -21.5 \text{ cal/mole-°K}.$$

For the reaction $Na(s) + \frac{1}{2}Cl_2(g) = NaCl(s)$, we thus find on the second convention:

$$\Delta\mathcal{S}^0_{298} = (\mathcal{S}^0_{298})_{NaCl} - [(\mathcal{S}^0_{298})_{Na} + \tfrac{1}{2}(\mathcal{S}^0_{298})_{Cl_2}]$$
$$= -21.5 - [0 + 0] = -21.5 \text{ cal/mole-°K}.$$

(c) The identity of the answers to (a) and (b) underlines the *purely* conventional nature of our choice of an entropy zero. On different conventions we of course obtain different S^0_T values for elements and compounds, but the figures for *any* ΔS^0_T remain exactly the same on all conventions. ◀

4

Consequences of the Thermodynamic Principles

The principles are the seed; now we reap some part of the harvest: the derivation and rationalization of a group of scientific laws that invest us with great powers of prediction. The point of departure for this development will be equation (23),

$$\Delta S \geq q/T,$$

where, remember, the equal sign refers to a reversible change and the inequality to an irreversible change.

Reversibility we conceived as an ideal limit, sometimes approached but never attained in any observable change. Proceeding under the impulsion of only infinitesimal differences in temperature, pressure, concentration, etc., a reversible change requires infinite time for its completion. But an observable change is, by definition, one that occurs in a finite (and usually quite brief) period. Thus some degree of irreversibility is necessarily associated with all changes actually occurring in the world of experience. Hence, for any observable change, $\Delta S > q/T$.

The better the approximation to reversibility attained in any particular change, the more nearly will q/T approach ΔS. In the limit we come to the reversible change proceeding through an endless sequence of states that differ only infinitesimally from true equilibrium states. But there can then be *no* empirically detectable difference between the infinitely enduring true equilibrium state and the state changing reversibly at an infinitesimal rate. As to conceptual difference(s), Caldin writes:

> Reversible change is characterized by an infinitesimal difference of intensity factors [pressure, temperature, electric potential, etc.] between system and surroundings: $I_{syst} - I_{surr} = \pm dI$. Equilibrium is defined by equality of these intensity factors: $I_{syst} - I_{surr} = 0$. These two equations are indistinguishable . . . *Any sufficient condition for reversibility is therefore also a sufficient condition for equilibrium.*

To define the equilibrium state we may then write $\Delta S = q/T$.

In the last two paragraphs we have discovered, in equation (23), criteria for the direction of spontaneous change and for the condition of equilibrium:

For an observable change $\Delta S > \frac{q}{T},$ (29)

For equilibrium $\Delta S = \frac{q}{T}.$ (30)

Though fundamental, these relations are not yet reduced to their most useful forms, from which are eliminated all terms that (like q) are not functions of state. Were chemical reactions conducted in isolated systems, optimal criteria for spontaneous change and equilibrium could at once be found in equation (24), $\Delta S \geq 0$. But chemical reactions are *not* run in isolated systems. Ordinarily they are run in systems held at *constant temperature* by contact with a thermostat; and held either at *constant pressure*, by contact with the atmosphere, or at *constant volume*, by enclosure within a rigid bomb. How can we best recast our criteria for application to these systems?

THE FREE ENERGIES

Consider a system capable only of pressure-volume work, and held at constant temperature and volume. Reuniting equations (29) and (30), for such a system we can write

$$\Delta S \geq q_V/T \qquad \text{or} \qquad 0 \geq q_V - T\,\Delta S.$$

But from equation (3) we know that $\Delta E = q_V$, so that

$$\Delta E - T\,\Delta S \leq 0. \tag{a}$$

For systems maintained at constant temperature and volume, we have thus contrived to recast, entirely in terms of functions of state, the fundamental criteria expressed in equations (29) and (30). And we can go one step further. Just as in similar circumstances we invented the enthalpy function, we now introduce a new function of state expressly designed for optimal expression of relation (a).

The Helmholtz free energy, or work function, is symbolized by the letter A (from Ger. *Arbeit*, work) and defined by the equation

$$A \equiv E - TS. \tag{31}$$

Defined entirely in terms of functions of state, A also is a function of state. Now compare two states of a system held at constant temperature:

$$A_2 = E_2 - TS_2,$$
$$A_1 = E_1 - TS_1.$$

Subtracting, we obtain an important general relation applicable to any *isothermal* change:

$$\Delta A = \Delta E - T\,\Delta S. \tag{32}$$

From (32) we can at once substitute in equation (a), where the equality and inequality signs still have the same connotations indicated in equations (29) and (30). For the special conditions under which (a) was derived, we thus obtain

The criterion of a spontaneous change at constant T and V:	$\Delta A < 0.$	(33)
The criterion of equilibrium in a system at constant T and V:	$\Delta A = 0.$	(34)

For a *reversible isothermal* change—with equation (2) assuming the form $\Delta E = q_{\text{rev}} - w_{\text{rev}}$, and $q_{\text{rev}} = T\,\Delta S$ by definition—substitution in equation (32) yields

$$\Delta A = (q_{\text{rev}} - w_{\text{rev}}) - q_{\text{rev}} = -w_{\text{rev}}$$

or

$$w_{\text{rev}} = -\Delta A. \tag{35}$$

Measured by the decline of a function of state (the "free energy"), the maximum work recoverable from a given isothermal change is thus shown to have a unique value, regardless of the path along which the change proceeds. Comparing (35) with the criteria expressed in equations (33) and (34), we observe that, under the indicated conditions, spontaneous change always reduces the system's capacity to perform work. And equilibrium is attained only when this capacity has been reduced to a minimum.

In the Helmholtz free energy we have discovered a function that can represent the changes and equilibria of chemical systems in much the same way that a potential-energy function represents change and equilibrium in purely mechanical systems. We see this most vividly when we prepare a graph showing, for any given reaction, the variation in the value of A with variation in the index of completion ξ. On the left, where $\xi = 0$, we plot the total value of A for all the reactants; on the right, where $\xi = 1$, we plot the total value of A for all the products; and in between we plot the total values of A for reaction mixtures represented by intermediate values of ξ. The curve obtained is usually of the type shown in Fig. 31. And the equilibrium state then lies at x where, in the trough of the curve, the condition $\Delta A = 0$ (or, better, $dA/d\xi = 0$) is at last satisfied when A assumes its minimum value.

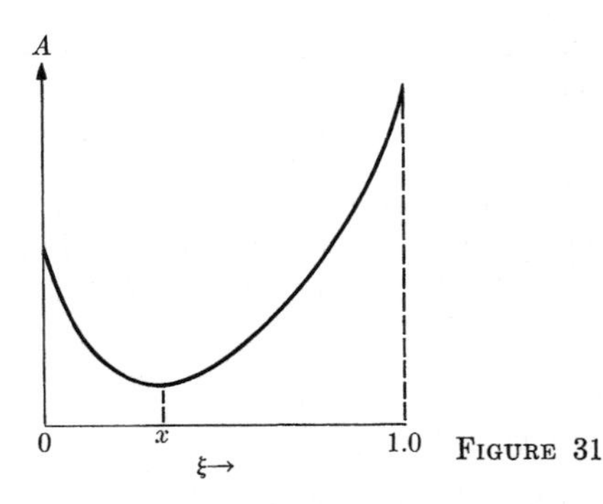

FIGURE 31

From any initial state to the final state of equilibrium, the progress of the isothermal reaction, in either direction, can now be viewed simply as a more or less swift "slide down to the bottom." The *maximum work output* recoverable from the spontaneous change is indicated by the vertical drop from the initial to the final state, i.e., $A_{\text{initial}} - A_{\text{final}} = -\Delta A = w_{\text{rev}}$. And a *minimum work input* of equal magnitude will be required to bring about the nonspontaneous inverse change (with $\Delta A > 0$) that restores the equilibrium system to its original nonequilibrium state. Since no real change ever proceeds strictly reversibly, the work output actually recovered from a spontaneous change is always somewhat less than the theoretical maximum represented by $-\Delta A$. But, as the difference in a function of state, $-\Delta A$ remains exactly the same however much or little work is recovered from the given change. Though the *total energy* of system and surroundings always remains constant, every spontaneous change—to the extent that it is irreversible—thus results in a "degradation of energy," in an irretrievable decrease in the amount of *free energy* recoverable as work.

Gibbs free energy

Exemplary for changes at constant temperature and *volume*, the criteria expressed in terms of ΔA do not fully meet the needs of chemists, who most often conduct their reactions under conditions of constant temperature and *pressure*. Under these conditions the fundamental criteria, for direction of change and condition of equilibrium, are better expressed in terms of the Gibbs free energy (G) defined by the equation

$$G \equiv H - TS. \tag{36}$$

Just as we obtained equation (32) from equation (31), for an *isothermal* change we find here

$$\Delta G = \Delta H - T\,\Delta S. \tag{37}$$

For a change proceeding at *constant pressure*, we can substitute for ΔH from equation (7), and thus obtain

$$\Delta G = \Delta E + P\,\Delta V - T\,\Delta S.$$

Substituting from equation (32), we have then

$$\Delta G = \Delta A + P\,\Delta V,$$

whence, in view of equation (35), we can conclude that

$$-\Delta G = w_{\text{rev}} - P\,\Delta V = w_{\text{net}}. \tag{38}$$

Here w_{net} represents the maximum recoverable work *exclusive* of pressure-volume work which is seldom recovered in chemical changes. Like w_{rev}, w_{net} is seen to be independent of the path along which the change proceeds under the constraints of *constant temperature and pressure*. In any actual change, the recovery of this "net work" will always fall short of $-\Delta G$, but $-\Delta G$ for the given change of state will of course remain the same, however much or little work we recover from it.

Let us now express in terms of G the fundamental criteria embodied in equations (29) and (30). Consider a system capable only of pressure-volume work, and held at constant temperature and pressure. Reuniting equations (29) and (30), for such a system we can write

$$\Delta S \geq q_P/T \qquad \text{or} \qquad 0 \geq q_P - T\,\Delta S.$$

But from equation (5) we know that $\Delta H = q_P$, so that

$$\Delta H - T\,\Delta S \leq 0. \tag{b}$$

Substituting now from equation (37), we find

The criterion of a spontaneous change at constant T and P: $\quad \Delta G < 0. \quad (39)$

The criterion of equilibrium in a system at constant T and P: $\quad \Delta G = 0. \quad (40)$

No system is at equilibrium if it can undergo a change that reduces its capacity to yield work. Equilibrium is attained only when the system attains the condition of minimum free energy specified by the criterion $\Delta G = 0$ or, better, $dG/d\xi = 0$.

The rôle of the Gibbs free energy as a potential function is nicely highlighted in its application to phase equilibria. Consider a pure substance that exists, at some constant temperature and pressure, in two interconvertible phases α and β (e.g., solid in contact with liquid, liquid in contact with vapor, etc.). How shall we formulate the condition for equilibrium between the two phases? We cannot simply insist on equality of the Gibbs free energies of the two phases: after all, equilibrium can obtain even when only a few milligrams of one phase stand in contact with a few tons of the other phase. But we can and do insist that at equilibrium the Gibbs free energy *per mole* of substance must be the same in both phases. Thus at equilibrium $\overline{G}_\alpha = \overline{G}_\beta$, and if this relation does *not* hold, the two-phase system is *not* at equilibrium. If for example $\overline{G}_\alpha > \overline{G}_\beta$, passage of material from α to β may proceed as a spontaneous process with $\Delta G < 0$, and the more stable phase β then grows at the expense of phase α—until either that phase is exhausted or the condition $\overline{G}_\alpha = \overline{G}_\beta$ is attained. Just so a difference of electrical potential evokes a passage of charge (nominally) from the region of high potential to any region of lower potential, and electrical equilibrium is established only when the electric potential is the same throughout the system. This is indeed the most familiar model for the conception of the molar free energy as a *chemical potential*—any difference of which determines the direction in which material is transferred, and equality of which is the essential characteristic of an equilibrium system.

We have just touched on phase equilibrium in one-component systems, but the concept of chemical potential is far more generally applicable. Indeed, it assumes its full importance mainly in *poly*component systems. Some indication of this larger sphere of usefulness will appear in our consideration of the familiar colligative properties. These we will find readily derivable from the proposition that, at equilibrium, the molar free energy of each component of a polycomponent system must be the same in every phase in which the component appears.* For were this relation ever to fail, the transfer of the component—from wherever its chemical potential is higher to wherever the potential is lower—could proceed with $\Delta G < 0$ as a spontaneous change in a system then demonstrably *not* at equilibrium.

Consider now the two terms that determine the equality or inequality of chemical potentials in a two-phase system containing one component.

---

* Where in a one-component system the chemical potential is a molar free energy ($\overline{G}$) defined by the quotient G/n, in a polycomponent system the chemical potential of the ith component is a partial molar free energy defined by the partial derivative $(dG/dn_i)_{T,P,n_a}, \ldots$ and usually symbolized μ_i. What μ_i represents is the free energy of one mole of the ith component, at a given temperature and pressure, in a solution of *given composition*. In determining μ_i we must then take care not to change that composition. In effect, the mathematical formulation instructs us to determine the infinitesimal change, in the total free energy of the solution, produced by an infinitesimal addition of the ith component—which addition should leave essentially unchanged the gross composition of the solution.

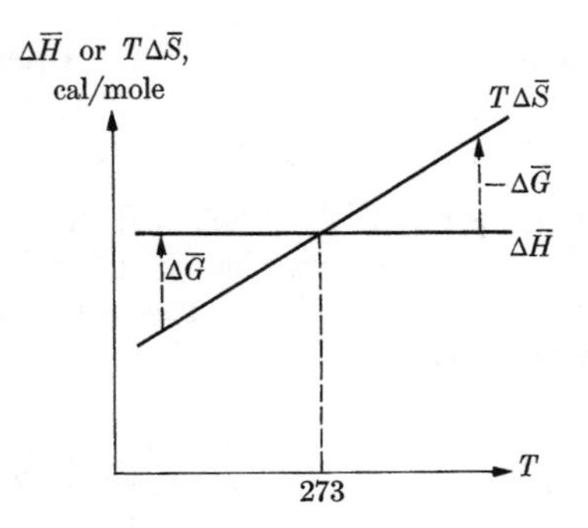

FIGURE 32

Taking as our example ice and water at the same temperature, we write

$$\overline{G}_w = \overline{H}_w - T\overline{S}_w,$$
$$\overline{G}_i = \overline{H}_i - T\overline{S}_i.$$

Subtracting the second equation from the first, we obtain

$$\Delta\overline{G} = \Delta\overline{H} - T\,\Delta\overline{S},$$

where this equation is here written with specific reference to the change:

$$H_2O\ (s) \rightarrow H_2O\ (\ell).$$

Now we know that, under 1 atm pressure, this change proceeds to the right at $T > 273°K$, to the left at $T < 273°K$, and that both phases stand in equilibrium at $T = 273°K$. This means that, for the indicated *isothermal* change, as it occurs at different temperatures,

$$\text{at } T > 273, \quad \Delta\overline{H} - T\,\Delta\overline{S} = \Delta\overline{G} < 0 \quad \text{or} \quad \Delta\overline{H} < T\,\Delta\overline{S};$$
$$\text{at } T = 273, \quad \Delta\overline{H} - T\,\Delta\overline{S} = \Delta\overline{G} = 0 \quad \text{or} \quad \Delta\overline{H} = T\,\Delta\overline{S};$$
$$\text{at } T < 273, \quad \Delta\overline{H} - T\,\Delta\overline{S} = \Delta\overline{G} > 0 \quad \text{or} \quad \Delta\overline{H} > T\,\Delta\overline{S}.$$

In the melting of ice both $\Delta\overline{H}$ and $\Delta\overline{S}$ are *positive* quantities, and both are approximately constant in the immediate vicinity of 273°K. Hence the situation is as represented in Fig. 32. Observe how at low temperatures the $\Delta\overline{H}$ term must *always* become dominant over a $T\,\Delta\overline{S}$ term that contains T as a multiplier. Thus T acts as a weighting factor which establishes the relative importance of $\Delta\overline{H}$ and $T\,\Delta\overline{S}$ terms, the balance between which determines the direction of spontaneous change and the position of equilibrium. And equilibrium will prevail just when, at the intersection of the two lines, $\Delta\overline{H} = T\,\Delta\overline{S}$. For it is precisely in a constant-pressure *equilibrium* process that we will be justified in writing

$$\Delta\overline{S} = q_{\text{rev}}/T = q_P/T = \Delta\overline{H}/T.$$

▶ *Example 1*

In the change

$$H_2O\ (\ell) \rightarrow H_2O\ (g,\ 1\ atm),$$

at 298°K and 1 atm pressure, $\Delta\overline{H} = 10.52$ kcal/mole and $\Delta\overline{S} = 28.39$ cal/mole-°K. (a) Is this a spontaneous change at 298°K? (b) Estimate the temperature at which these liquid and gaseous phases would stand in equilibrium with each other.

Solution

$$\text{(a)}\qquad \Delta\overline{G}_{298} = \Delta\overline{H}_{298} - T\,\Delta\overline{S}_{298} = 10{,}520 - 298(28.39)$$
$$= 10{,}520 - 8{,}460 = +2{,}060 \text{ cal/mole} > 0.$$

At 298°K the favored change is thus not the indicated evaporation but, rather, the *inverse* process in which $\Delta\overline{G} = -2{,}060$ cal/mole attends the condensation to liquid of water vapor at 1 atm pressure.

(b) When the two phases stand in equilibrium, $\Delta\overline{G} = 0$ and $\Delta\overline{H} = T\,\Delta\overline{S}$. Assuming at least *approximate* constancy of $\Delta\overline{H}$ and $\Delta\overline{S}$ over the temperature range concerned, we write

$$T_{eq} = \Delta\overline{H}/\Delta\overline{S} = \frac{10{,}520}{28.39} = 371°\text{K}.$$

The correct answer would obviously be the normal boiling point of water, 373.15°K. Hence the approximation is not a bad one—due mainly to mutual cancellation of the effects of quite substantial changes actually occurring in the values of both $\Delta\overline{H}$ and $\Delta\overline{S}$. You should be able to see why some error-reducing cancellation of this sort will occur whenever, as here, $\Delta\overline{H}$ and $\Delta\overline{S}$ are of the *same* sign. ◀

Turning now from phase equilibria to chemical equilibria, we still find in the equation $\Delta G = \Delta H - T\,\Delta S$ an indicator of how the favored direction of reaction is determined. Evidently decrease of enthalpy ($\Delta H < 0$) and increase of entropy ($\Delta S > 0$) both act to favor spontaneous reaction ($\Delta G < 0$). But this is only the first of four possible situations displayed in Table 4. The second is simply the inverse of the first: when ΔH is positive and ΔS negative, ΔG will be positive at all temperatures, and at *no* temperature will the reaction proceed spontaneously. The last two possibilities, in which ΔH and ΔS are alike in sign, are those in which opposite directions of reaction are favored at high and low temperatures. At sufficiently low temperatures, minimizing the $T\,\Delta S$ term, the ΔH term will be dominant; and the favored direction of reaction will be that in which change is exothermic (negative ΔH). At sufficiently high temperatures, however, the $T\,\Delta S$ term must become dominant, and the favored direction of reaction will then be that in which change proceeds with increase of entropy

TABLE 4

ΔH	ΔS	ΔG
$-$	$+$	$-$, reaction always favored
$+$	$-$	$+$, reaction never spontaneous
$-$	$-$	?, reaction favored at low temperature
$+$	$+$	?, reaction favored at high temperature

(positive ΔS).[32] These are powerful generalizations but, like other thermodynamic propositions, they refer only to the equilibrium condition of the reaction concerned, and tell us nothing whatsoever about the rate at which that condition is attained. Remember too that predictions based on fixed values of ΔH and ΔS will become unreliable when, over greatly extended ranges of temperature, these values may change not only in magnitude but even in sign.

Our association of entropy with "spread" now suggests some rough but useful rules of thumb. Proceeding with minimal change in spread, reactions that involve only condensed phases (i.e., liquids or solids), as well as reactions in which the same number of moles of gas figure in reactants and products, will most often be reactions with comparatively small values of ΔS. If either direction of such a reaction is substantially exothermic, we will usually be right in predicting this as the favored direction at all save *very* high temperatures. On the other hand, if the reaction is one in which the number of moles of gas changes, to this change of spread should correspond a comparatively large value of ΔS. The reaction that *increases* the number of moles of gas present will then be the favored direction of reaction at all reasonably high temperatures, and even at reasonably low temperatures too if the reaction is not a strongly exothermic one.[33] To all such rules of thumb there are of course significant exceptions. Thus even when only condensed phases are involved, the formation of the highly ordered structure of a complex compound will be a change understandably associated with a large negative value of ΔS. If the synthetic reaction is sufficiently exothermic, the formation of the compound may well be favored at low temperatures. But as the temperature is raised, this compound, like complex compounds generally, will soon become unstable as the $T\,\Delta S$ term takes command.

Evaluations of free energies

As in the earlier instances of H and S, only *differences* of G are thermodynamically significant. But, just as in those earlier cases, for convenience we seek to construct a scale of free energies on which a distinct value can be established for each particular state of any given system. Again we proceed by defining a group of reference states to which free energies are assigned by convention. To the elements, in their stablest forms at 298°K and 1 atm pressure, we assign $G^0_{298} \equiv 0$. A straightforward calculation then yields the free energy of formation (ΔG^0_f) of one mole of any pure compound in its standard state at 298°K. How? Observe that

$$\Delta G^0_f = \Delta H^0_f - T\,\Delta S^0_f.$$

We have already shown that a standard heat of formation (ΔH^0_f) is determinable—either directly, or indirectly with the aid of Hess's law. We have also shown how, from purely thermal data and the Nernst heat theorem, values of S^0_{298} are attainable for all pure elements and compounds. Given these, we readily obtain a value of ΔS^0_f that Example 8 (p. 89) shows to be wholly independent of the fact that we set our entropy-zero not at 298°K but at 0°K. And then, knowing both ΔS^0_f and ΔH^0_f, we easily find ΔG^0_f.

Since G is a function of state, figures for ΔG can be added, subtracted, and manipulated in the very same way that figures for ΔH are handled in Hess-law calculations. Hence even when ΔH^0_f and ΔS^0_f values for some particular compound are unavailable, we may still be able to determine ΔG^0_f for that compound by an indirect calculation in the style of Hess's law. Moreover, once in possession of all relevant figures for ΔG^0_f (at 298°K), we can easily determine ΔG^0_{298} for any reaction—just as, given figures for ΔH^0_f, we earlier calculated ΔH^0_{298}. To express the general case, we symbolize by n_R the number of moles of each reactant (R) consumed, and by n_P the number of moles of each product (P) formed. For the overall reaction, as written, the value of ΔG^0_{298} is then given by the equation:

$$\Delta G^0_{298} = \sum n_P(\Delta G^0_f)_P - \sum n_R(\Delta G^0_f)_R.$$

▶ *Example 2*

The following data apply at 298°K and 1 atm.

	ΔH^0_f, kcal/mole	S^0_{298}, cal/mole-°K
$CuBr_2$ (s)	−33.2	30.1
$CuBr$ (s)	−25.1	21.9
Br_2 (g, 1 atm)	+ 7.34	58.64

Consider the reaction: $CuBr_2$ (s) = $CuBr$ (s) + $\frac{1}{2}$ Br_2 (g, 1 atm). (a) In what direction does this reaction proceed at 298°K and 1 atm pressure? (b) At what temperature will the three substances coexist at equilibrium under a pressure of 1 atm?

Solution. (a) The favored direction of reaction will be indicated by the sign of ΔG^0_{298}, to calculate which we need only determine ΔH^0_{298} and ΔS^0_{298} for the reaction.

$$\begin{aligned}\Delta H^0_{298} &= (\Delta H^0_f)_{\text{CuBr}} + \tfrac{1}{2}(\Delta H^0_f)_{\text{Br}_2} - (\Delta H^0_f)_{\text{CuBr}_2} \\ &= -25.1 + 3.67 - (-33.2) = +11.8 \text{ kcal/mole.}\end{aligned}$$

$$\begin{aligned}\Delta S^0_{298} &= (S^0_{298})_{\text{CuBr}} + \tfrac{1}{2}(S^0_{298})_{\text{Br}_2} - (S^0_{298})_{\text{CuBr}_2} \\ &= +21.9 + 29.32 - 30.1 = +21.1 \text{ cal/mole-°K.}\end{aligned}$$

$$\begin{aligned}\Delta G^0_{298} = \Delta H^0_{298} - T\,\Delta S^0_{298} &= 11{,}800 - (298)(21.1) \\ &= 5500 \text{ cal/mole} = 5.5 \text{ kcal/mole.}\end{aligned}$$

Not rightward but, rather, *leftward* progress of the reaction is thus strongly favored at 298°K, and will reduce the Br_2 pressure far below 1 atm.

(b) At whatever temperature (T) the three substances coexist at 1 atm pressure, we will have $\Delta H^0_T - T\,\Delta S^0_T = \Delta G^0_T = 0$. Noting that ΔH^0 and ΔS^0 are the same in sign, we essay the approximation in which the two terms are treated as substantially constant over the temperature range concerned. In that case:

$$\Delta H^0_T \cong \Delta H^0_{298} = 11.8 \text{ kcal/mole;}$$
$$\Delta S^0_T \cong \Delta S^0_{298} = 21.1 \text{ cal/mole-°K.}$$

And then

$$11{,}800 - T(21.1) = \Delta G^0_T = 0,$$
$$T = 11{,}800/21.1 = 560\text{°K.}$$

Our approximation has been stretched over a temperature range $<300°$, and problem 46 indicates that the approximation is actually quite good. ◀

Where the temperature dependence of ΔH^0 and ΔS^0 arises only from ΔC_P-terms that are usually small, ΔG^0 shows a very strong temperature dependence arising from the explicit presence of T in the very definition of G. We must now consider how to evaluate this dependence. And we ought at the same time to consider the significant variation of ΔG with pressure: only when we know how to allow for this variation would we be in a position to solve the last example for the temperature at which the equilibrium pressure of Br_2 (g) has some particular value *other than* 1 atm. Seeking a functional expression for the dependence of ΔG on P and T, we begin by constructing a powerful new combination of the first and second principles.

For a given change, proceeding reversibly along a given path, the first principle assumes the form:

$$dE = q_{\text{rev}} - w_{\text{rev}}.$$

Recalling how dS was defined in terms of q_{rev}, and stipulating that only pressure-volume work shall be possible, we obtain

$$dE = T\,dS - P\,dV. \tag{41}$$

Derived here in the special case of a reversible change, this important equation is easily shown to be more generally applicable. For imagine that the same change were to proceed along the same path in a partially or totally irreversible manner. The work then recovered (w) would be less than w_{rev} by some margin we symbolize as δ in the equation $w = w_{rev} - \delta$. But in the irreversible change the heat term (q) correspondent to w must be a very similar function of δ. For whether the given change proceeds reversibly or irreversibly, the value of dE must remain exactly the same. This invariance of dE will be ensured only if $q = q_{rev} - \delta$, so that in the irreversible change,

$$dE = q - w = (q_{rev} - \delta) - (w_{rev} - \delta) = q_{rev} - w_{rev} = T\,dS - P\,dV.$$

Thus equation (41) has been shown to apply to irreversible as to reversible changes.*

Turning now to the definition of G, we draw also on the definition of H to write

$$G \equiv H - TS \equiv E + PV - TS.$$

Differentiation yields

$$dG = dE + P\,dV + V\,dP - T\,dS - S\,dT.$$

Substituting for dE from equation (41), we find

$$dG = V\,dP - S\,dT. \tag{43}$$

ALL our subsequent work will take departure from this equation.

For one mole of material held at constant pressure, with $dP = 0$ equation (43) is reduced to $(d\overline{G}/dT)_P = -\overline{S}$. Since $\overline{S}$ is always a positive quantity, we learn from this equation that $\overline{G}$ must always *decrease* with rise of temperature. And this decrease will be particularly conspicuous with gases, which show comparatively large values of $\overline{S}$. For one mole of material held at constant temperature, with $dT = 0$ equation (43) is reduced to $(d\overline{G}/dP)_T = \overline{V}$. Since $\overline{V}$ is always a positive quantity, $\overline{G}$ must always *increase* with rise of pressure. Save for extremely large pressure

* Problem 42 invites derivation of two other equations of very similar form:

$$\left.\begin{aligned} dH &= T\,dS + V\,dP \\ dA &= -P\,dV - S\,dT \end{aligned}\right\}. \tag{42}$$

changes, this increase will generally be significant only with gases, which show conspicuously large values of $\overline{V}$.

Having thus established how changes in P and T affect $\overline{G}$ for a given pure substance, now we ask how such changes affect ΔG for a given reaction. For all the reaction products, *collectively*, we can write equation (43) as

$$dG_{\text{prod}} = V_{\text{prod}}\, dP - S_{\text{prod}}\, dT.$$

For all the reactants, *collectively*,

$$dG_{\text{reac}} = V_{\text{reac}}\, dP - S_{\text{reac}}\, dT.$$

Subtracting the second equation from the first, we arrive at

$$\begin{aligned} dG_{\text{prod}} - dG_{\text{reac}} &= d(G_{\text{prod}} - G_{\text{reac}}) \\ &= (V_{\text{prod}} - V_{\text{reac}})\, dP - (S_{\text{prod}} - S_{\text{reac}})\, dT. \end{aligned}$$

But these differences can be expressed far more compactly, as

$$d(\Delta G) = \Delta V\, dP - \Delta S\, dT, \tag{44}$$

where each Δ-term refers to the change in the indicated extensive property, consequent to the reaction. And so we have attained what we sought: a functional expression from which, given values of ΔV and ΔS, we can calculate how ΔG varies with temperature and pressure.

▶ *Example 3*

(a) Is diamond or graphite the more stable form of carbon at 298°K and 1 atm pressure? (b) At 298°K, what pressure would be required to form diamond (density, 3.5 gm/ml) from graphite (density, 2.25 gm/ml)? (c) How might one hope to gain some evidence for the soundness of this entire mode of analysis? Some relevant data are as follows:

$$\begin{aligned} &\text{Heat capacity data yield for diamond } \overline{S}^0_{298} = 0.58 \text{ cal/mole-°K} \\ &\text{Heat capacity data yield for graphite } \overline{S}^0_{298} = 1.37 \text{ cal/mole-°K} \\ \hline &\text{For C (graphite)} \rightarrow \text{C (diamond)}: \Delta\overline{S}^0_{298} = -0.79 \text{ cal/mole-°K} \end{aligned}$$

$$\begin{aligned} &\text{On combustion: C (graph)} + O_2 = CO_2,\ \Delta\overline{H}^0_{298} = -94.03 \text{ kcal} \\ &\text{On combustion: C (diam)} + O_2 = CO_2,\ \Delta\overline{H}^0_{298} = -94.48 \text{ kcal} \\ \hline &\text{For C (graphite)} \rightarrow \text{C (diamond)}: \Delta\overline{H}^0_{298} = +450 \text{ cal/mole} \end{aligned}$$

And the given densities imply:

$$\begin{aligned} &\text{For diamond: gm-atomic volume} = 12/3.5 = 3.4 \text{ ml/gm-atom} \\ &\text{For graphite: gm-atomic volume} = 12/2.25 = 5.3 \text{ ml/gm-atom} \\ \hline &\text{For C (graphite)} \rightarrow \text{C (diamond)}: \Delta\overline{V}_{298} = -1.9 \text{ ml/gm-atom} \end{aligned}$$

Solution. (a) For the reaction C (graphite) $\rightarrow$ C (diamond),

$$\Delta\overline{G}^0_{298} = \Delta\overline{H}^0_{298} - T\,\Delta\overline{S}^0_{298} = 450 - 298(-0.79) = +685 \text{ cal/mole.}$$

At room temperature and atmospheric pressure, the favored direction of reaction is thus the *inverse* change of diamond into graphite. Under ordinary conditions diamond is therefore a thermodynamically unstable species, which exists only because of the extreme slowness of its conversion into the more stable graphite.

(b) At atmospheric pressure the reaction C (graphite) $\rightarrow$ C (diamond) is characterized by $\Delta\overline{G}^0_{298} > 0$. But with $\Delta\overline{V} < 0$ in this reaction, a sufficient rise of pressure should change the sign of $\Delta\overline{G}$ and, hence, the favored direction of reaction at 298°K. For consider that the effect of a change of pressure at *constant temperature* will be given by the following reduced form of equation (44):

$$d(\Delta\overline{G}) = \Delta\overline{V}\,dP.$$

If we approximate by treating $\Delta\overline{V}$ as constant ($= -0.0019$ lit) over the entire pressure range involved, the last equation can easily be integrated. For the upper limit we choose that pressure (P^*) at which $\Delta\overline{G}^0_{298} = 0$, i.e., the pressure under which diamond and graphite stand in equilibrium with each other at 298°K. For the lower limit we use the standard pressure of 1 atm, at which we have just found $\Delta\overline{G}^0_{298} = +685$ cal/mole. However, with $\Delta\overline{V}$ in liters and pressure in atmospheres, the appropriate unit for $\Delta\overline{G}$ is not calories but lit-atm. Multiplication by the factor 0.0413 converts a-figure in calories to one in lit-atm, so that $\Delta\overline{G}^0_{298} = 0.0413(685) =$ 28.3 lit atm. We have then

$$\int_{28.3}^{0} d(\Delta\overline{G}) = -0.0019\int_{1}^{P^*} dP,$$

$$0 - 28.3 = -0.0019(P^* - 1),$$

$$P^* \cong 15{,}000 \text{ atm.}$$

At 25°C diamond and graphite would stand in equilibrium under a pressure of 15,000 atm. At still higher pressures graphite becomes thermodynamically unstable, and its conversion into diamond is then possible in principle, though so slow in practice as to be wholly undetectable.

(c) Since at room temperature no equilibrium of graphite with diamond ever *is* attained in practice, there remains room for skepticism that we have correctly calculated what *would* be the equilibrium condition. But, given expressions for $\Delta\overline{H}$ and $\Delta\overline{V}$ as functions of temperature and pressure, by integration of equation (44) we can calculate that the equilibrium pressure is of the order of 75,000 atm at 1500°K. And *here* the soundness of our calculation is attested by an unmistakable production of diamond from graphite at pressures that exceed the equilibrium pressure! ◀

Gibbs-Helmholtz equations

By variously combining equations (36) and (43) or equations (37) and (44)—and from similar combinations of the corresponding equations in terms of A rather than G—one obtains a group of so-called Gibbs-Helmholtz relations. Evidently such a relation can contain nothing not already implicit in the equations combined in making the derivation, but the Gibbs-Helmholtz relations often prove to be remarkably *convenient* formulations. To produce one representative example of a Gibbs-Helmholtz equation, we begin by recalling that for a constant-pressure change equation (44) is reduced to

$$\left[\frac{d(\Delta G)}{dT}\right]_P = -\Delta S.$$

Substituting in equation (37) this expression for $-\Delta S$, we find

$$\Delta G = \Delta H + T\left[\frac{d(\Delta G)}{dT}\right]_P.$$

To recast this in a more compact form, we divide through by $-T^2$, and rearrange to obtain

$$-\frac{\Delta G}{T^2} + \frac{1}{T}\left[\frac{d(\Delta G)}{dT}\right]_P = -\frac{\Delta H}{T^2},$$

$$\Delta G\,\frac{d}{dT}\left[\frac{1}{T}\right]_P + \frac{1}{T}\,\frac{d}{dT}\,[\Delta G]_P = -\frac{\Delta H}{T^2}.$$

Considering now that $u\,dv + v\,du = d(uv)$, we conclude that

$$\frac{d}{dT}\left[\frac{\Delta G}{T}\right]_P = -\frac{\Delta H}{T^2}.$$

We will put this equation to good use in two subsequent developments.

THE CLAPEYRON EQUATION

Consider a pure substance that exists in two phases (denoted by subscripts 1 and 2) which stand in equilibrium with each other at some particular temperature and pressure. We have found as the essential condition for such an equilibrium

$$\overline{G}_1 = \overline{G}_2.$$

Imagine now a further equilibrium state established by making appropriate infinitesimal changes (dT and dP) in the original temperature and pressure. Since the new state is again an equilibrium state, satisfaction of the same essential condition now entails that

$$\overline{G}_1 + d\overline{G}_1 = \overline{G}_2 + d\overline{G}_2.$$

Together, the last two equations yield

$$d\overline{G}_1 = d\overline{G}_2.$$

To establish the relation of dT to dP in the change that links the two equilibrium states, we have only to substitute from equation (43).

$$\overline{V}_1\, dP - \overline{S}_1\, dT = \overline{V}_2\, dP - \overline{S}_2\, dT,$$
$$(\overline{S}_2 - \overline{S}_1)\, dT = (\overline{V}_2 - \overline{V}_1)\, dP,$$
$$dT/dP = \Delta\overline{V}/\Delta\overline{S}, \tag{c}$$

where $\Delta\overline{S}$ and $\Delta\overline{V}$ denote the changes of entropy and volume, respectively, when one mole of phase 1 is converted into phase 2.*

What is the order of magnitude of dT/dP in some familiar changes? Near room temperature a representative vaporization proceeds with $\Delta\overline{V} \cong 22{,}000$ ml/mole and (by Trouton's rule) $\Delta\overline{S} \cong 21$ cal/mole-°K. If, with $\Delta\overline{V}$ expressed in ml, we seek an expression of dT/dP in °K/atm, we must multiply by the factor 41.3 ml-atm/cal to convert $\Delta\overline{S}$ to the units ml-atm/mole-°K. For only then will the dimensions check out, as they do here:

$$\frac{dT(°\text{K})}{dP(\text{atm})} = \frac{\Delta\overline{V}\left(\dfrac{\text{ml}}{\text{mole}}\right)}{\Delta\overline{S}\left(\dfrac{\text{ml-atm}}{\text{mole-°K}}\right)}.$$

Now in the vaporization process

$$\frac{dT}{dP} = \frac{22{,}000}{21(41.3)} \cong +25°\text{K/atm}.$$

Even a small change in ambient pressure will thus significantly change the temperature at which a liquid stands at equilibrium with its vapor, and in measuring the normal boiling point of a liquid we must always take care that the pressure does not deviate significantly from 1 atm. Far different is the situation when only condensed phases are involved, so that $\Delta\overline{V}$ is

---

* Observe that the last equation is even more easily derivable from equation (44). For *both* the original and final equilibrium states, we know that $\Delta\overline{G} = 0$. The infinitesimal changes dT and dP, which link these equilibrium states, thus leave $\Delta\overline{G}$ wholly unaltered. Hence $d(\Delta\overline{G}) = 0$, and equation (44) becomes

$$\Delta\overline{V}\, dP - \Delta\overline{S}\, dT = d(\Delta\overline{G}) = 0,$$

which at once yields equation (c) above.

comparatively small. In a representative melting process we might find $\Delta\overline{S} \cong 3$ cal/mole-°K and $\Delta\overline{V} \cong 6$ ml/mole, and in that case:

$$\frac{dT}{dP} = \frac{6}{3(41.3)} \cong +0.05\text{°K/atm.}$$

Change of pressure thus has relatively little effect on the temperature at which a solid stands at equilibrium with its melt, and in measuring the normal melting point of a solid we ordinarily pay no attention to small deviations of ambient pressure from 1 atm.

At equilibrium, when $\Delta\overline{G} = 0$, equation (37) requires that $\Delta\overline{S} = \Delta\overline{H}/T$. Substitution for $\Delta\overline{S}$ in equation (c) then yields

$$\frac{dT}{dP} = \frac{T\,\Delta\overline{V}}{\Delta\overline{H}}, \tag{45}$$

which is perhaps the most familiar form of the Clapeyron equation. Combining rigor with extreme generality, this simple relation applies alike to vaporization, fusion, and sublimation equilibria—as well as to equilibria of two allotropic forms like diamond and graphite. Given expressions for $\Delta\overline{H}$ and $\Delta\overline{V}$, and knowing beside only the temperature and pressure prevailing at some *one* particular equilibrium state, we can integrate the Clapeyron equation to obtain the entire set of coordinated temperatures and pressures that describe *all* equilibrium states of the system. Conversely, given actual measurements of dT/dP and $\Delta\overline{V}$ for some phase transition, we can calculate for that change a value of $\Delta\overline{H}$ that might otherwise be hard to determine (e.g., under conditions of extreme pressure).

► *Example 4*

Ice and water stand in equilibrium with each other under a pressure of 1 atm at 273.15°K. (a) Given 0.917 gm/ml and 1.000 gm/ml as the densities of ice and water respectively, and 80 cal/gm as the heat of fusion of ice, calculate dT/dP in the vicinity of 273°K. (b) What part of the result just obtained is derivable from Le Chatelier's principle, and what part is not so derivable? (c) In calculations employing the Clapeyron equation, is it really necessary to use the *molar* values of ΔH and ΔV?

Solution. (a) Taking 18 as the molecular weight of water, we formulate as follows the change of molar volume in the conversion of ice to water:

$$\Delta\overline{V} = \frac{18}{1.00} - \frac{18}{0.917} = 18(1 - 1.091) = 18(-0.091)\text{ ml/mole.}$$

When ice melts, the molar change of enthalpy is $\Delta\overline{H} = 18(80)$ cal/mole or 18(80)(41.3) ml-atm/mole. Substituting in equation (45), we find

$$\frac{dT}{dP} = \frac{(273)(18)(-0.091)}{(18)(80)(41.3)} = -\frac{(273)(0.091)}{(80)(41.3)} = -0.0075°\text{K/atm.}$$

This result has *some* bearing on the relation of the triple point to the normal freezing point of water.[34]

(b) Since ice is less dense or more voluminous than water, Le Chatelier's principle suggests that increase of pressure should promote the melting of ice. To maintain the equilibrium of ice and water under increased pressure, some reduction of temperature will then be required. From Le Chatelier's principle we thus correctly infer that dT/dP is *negative*. But this principle is incapable of supplying the precise numerical value of dT/dP we have derived from our thermodynamic analysis. And this is everywhere the superior relationship in which quantitative thermodynamics stands to the purely qualitative Le Chatelier principle.[35]

(c) In the above calculation the cancellation of the molecular weight of water (18) makes evident what may already have been guessed: ΔH and ΔV need only refer to the *same* quantity of material (e.g., one gram) and not necessarily to one mole. Observe too that it makes no difference whether ΔH and ΔV refer to the melting of ice or to the freezing of water. An answer correct in sign (and magnitude) is obtained in either case, provided only that ΔH and ΔV refer to the *same* direction of change. ◀

The Clausius-Clapeyron equation

We can recast the Clapeyron equation in a convenient approximate form applicable to vaporization and sublimation equilibria, in which one of the two phases is gaseous. Let us approximate first by supposing that the molar volume of the condensed phase ($\overline{V}_c$) can be neglected in comparison with the molar volume of the corresponding gas phase ($\overline{V}_g$). As a further approximation, we treat the gas as though it were ideal. Then

$$\Delta\overline{V} = \overline{V}_g - \overline{V}_c \cong \overline{V}_g \cong \frac{RT}{P}\cdot$$

Substitution for $\Delta\overline{V}$ in equation (45) then yields

$$\frac{dT}{dP} = \frac{RT^2}{P\,\Delta\overline{H}}\cdot$$

Inverting and rearranging, we arrive at the following useful form of the Clausius-Clapeyron equation

$$\frac{dP}{P} = d\ln P = \frac{\Delta\overline{H}\,dT}{RT^2}\cdot \tag{46}$$

As a third and final simplifying approximation, we treat $\Delta\overline{H}$ as though it were a constant independent of temperature. Integration of the last equation then becomes trivially easy:

$$\int_{P_1}^{P_2} d\ln P = \frac{\Delta\overline{H}}{R}\int_{T_1}^{T_2}\frac{dT}{T^2},$$

$$\ln\frac{P_2}{P_1} = -\frac{\Delta\overline{H}}{R}\left(\frac{1}{T_2}-\frac{1}{T_1}\right). \tag{47}$$

Despite the several approximations involved, equations (46) and (47) apply excellently to an immense number of measurements of vapor pressure over various liquids and solids. Plots of $\ln P$ *vs* $1/T$ are generally quite satisfactorily rectilinear, and from the slope of such a plot ($= -\Delta\overline{H}/R$) the molar heat of volatilization can be read off with surprising accuracy.[36]

IDEAL SOLUTIONS AND COLLIGATIVE PROPERTIES

In the colligative properties of ideal solutions we encounter some new kinds of phase equilibria. These include the equilibrium of a solution with pure solvent vapor (as manifested in vapor-pressure lowering and boiling-point elevation), the equilibrium of solution with pure solid solvent (as manifested in freezing-point depression), and the equilibrium of solution with pure liquid solvent (as manifested in pressure difference across an osmotic membrane). To these we again apply our most powerful criterion of equilibrium: in the particular pair of phases concerned in each case, the material common to both phases must have the same molar free energy. But where before we were concerned with the equilibrium of phases containing only a single pure component, we must now for the first time consider how the molar free energy of a dissolved substance depends on its concentration in the solution.

The dependence of molar free energies on temperature and pressure we have determined from equation (43), $d\overline{G} = \overline{V}\,dP - \overline{S}\,dT$. This equation was derived on the tacit assumption that two parameters suffice to define the state of a phase—as is indeed the case when the phase contains a single pure substance, or at least a material invariant in its composition. But two parameters are *not* enough fully to define the state of, say, a two-component solution: beyond temperature and pressure, the concentration of the solution must now also be specified. When we have to deal with solutions, then, only *part* of what we need to know is given us by equation (43), the limitation of which we can make explicit by writing

$$d\overline{G}_X = \overline{V}\,dP - \overline{S}\,dT.$$

Here the subscript X indicates constancy of composition, i.e., mole fractions. Under conditions of constant composition and constant temperature, the pressure dependence of the molar free energy will then be written

$$d\overline{G}_{XT} = \overline{V}\,dP.$$

Under conditions of constant composition and constant pressure, the temperature dependence of the molar free energy will be written

$$d\overline{G}_{XP} = -\overline{S}\,dT.$$

And now, going beyond equation (43), we seek an expression that, for a component in a solution held at constant temperature and pressure, will represent the change of free energy ($d\overline{G}_{TP}$) associated with a change of concentration (dX). That is, we seek the function $f(X)$ that will permit us to write:

$$d\overline{G}_{TP} = f(X)\,dX.$$

With a view to the colligative properties that will concern us, we begin by considering a very dilute solution of an involatile solute in a volatile solvent. Whenever solvent vapor, at some particular pressure, stands in equilibrium with liquid solvent, at some particular concentration in the solution, we know that

$$\overline{G}_{\ell} = \overline{G}_{g}.$$

Here $\overline{G}_{\ell}$ and $\overline{G}_{g}$ symbolize the molar free energies of solvent in solution and vapor phases respectively. *Without change of temperature*, let a new condition of equilibrium be produced by appropriate infinitesimal alterations in the concentration of the liquid solvent and the pressure of the gaseous solvent. For this new equilibrium state we will have

$$\overline{G}_{\ell} + d\overline{G}_{\ell} = \overline{G}_{g} + d\overline{G}_{g}.$$

Together the last two equations yield

$$d\overline{G}_{\ell} = d\overline{G}_{g}.$$

In the gas phase, containing only pure solvent vapor, the change in molar free energy ($d\overline{G}_{g}$) arises solely from the alteration of pressure. For this isothermal alteration, in a system of constant composition, equation (43) yields

$$d\overline{G}_{g} = \overline{V}_{g}\,dP,$$

where $\overline{V}_{g}$ symbolizes the molar volume of the solvent vapor.

In the liquid phase the situation is somewhat more complex. Here there have been alterations both in the concentration of the liquid solvent and

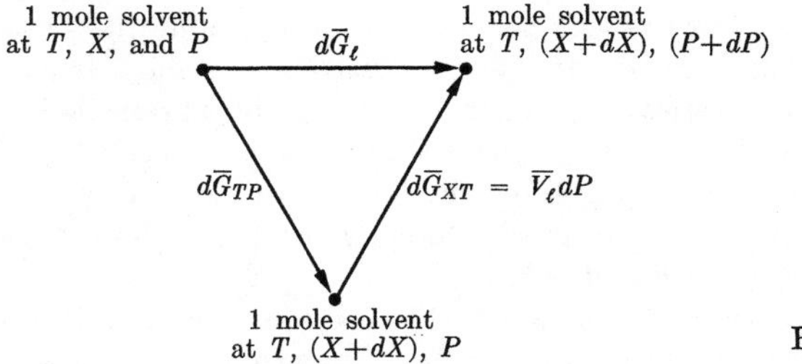

FIGURE 33

in the pressure exerted on the solution by the superincumbent vapor. How shall we calculate the *net* change ($d\overline{G}_\ell$) in the molar free energy of the liquid solvent? Considering that G is a function of state, we need only call again on the familiar expedient sketched in Fig. 33. For the isothermal change of both concentration and pressure, $d\overline{G}_\ell$ can be equated to the *sum* of $d\overline{G}_{TP}$, reflecting the isothermal change of concentration only, *plus* $d\overline{G}_{XT}$, reflecting the isothermal change of pressure only. Again equation (43) permits us to express the effect of the pressure change as $d\overline{G}_{XT} = \overline{V}_\ell\, dP$, where $\overline{V}_\ell$ symbolizes the molar volume of the solvent in the liquid phase. Equating the changes along the one- and two-step paths joining the same initial and final states, we find

$$d\overline{G}_\ell = d\overline{G}_{TP} + \overline{V}_\ell\, dP.$$

Substitution in the equilibrium condition, $d\overline{G}_\ell = d\overline{G}_g$, then yields

$$d\overline{G}_{TP} + \overline{V}_\ell\, dP = \overline{V}_g\, dP$$

$$d\overline{G}_{TP} = (\overline{V}_g - \overline{V}_\ell)\, dP.$$

We have thus attained an analytical expression for $d\overline{G}_{TP}$, but this is not yet a very useful expression. We can push a step further by making the same double approximation used in deriving the Clausius-Clapeyron relation. Regarding the molar volume of the liquid as negligible compared to the molar volume of the vapor, we further suppose the vapor to be ideal. For the change in the molar free energy of the solvent in the solution, we have then

$$d\overline{G}_{TP} = (\overline{V}_g - \overline{V}_\ell)\, dP \cong \overline{V}_g\, dP \cong \frac{RT}{P}\, dP = RT\, d \ln P. \qquad \text{(d)}$$

This is a rather curious result. We have sought an expression for the change $d\overline{G}_{TP}$, arising from alteration of the solvent's concentration in a solution held at constant temperature and pressure. And we have found an expression for $d\overline{G}_{TP}$—but only in terms of the change in the pressure, of the pure solvent vapor, associated with that same alteration in concentration. Referring as it does to the change in the molar free energy of the liquid solvent, surely $d\overline{G}_{TP}$ ought to be directly expressible in terms of the alteration of the solvent's concentration in the solution. We attain such an expression by forging a new link between the liquid and vapor phases. We stipulate that in an *ideal* solution the solvent must conform at least approximately to Raoult's law, $P = P^0X$. Here the vapor pressure (P) of solvent over a solution is linked to its mole fraction (X) in the solution by a proportionality constant (P^0) representing the vapor pressure of the pure solvent at the same temperature. In logarithmic form Raoult's law reads

$$\ln P = \ln P^0 + \ln X.$$

Under conditions of constant temperature and pressure, the vapor pressure of the pure solvent is of course a constant, so that differentiation of the last equation will yield

$$d \ln P = d \ln X.$$

Substituting now in the last equation of the preceding paragraph, we arrive at

$$d\overline{G}_{TP} = RT\, d \ln X. \tag{48}$$

This is just what we sought. The change in the molar free energy of the liquid solvent has now been expressed in terms of the change in its concentration in an ideal solution maintained at constant temperature and pressure. Integrating the last equation, we take as the lower limits $X = 1$ for the pure liquid solvent with molar free energy $\overline{G}_\ell^0$, and as upper limits the mole fraction X at which the molar free energy of the solvent in the solution is $\overline{G}_\ell$.

$$\int_{\overline{G}_\ell^0}^{\overline{G}_\ell} d\overline{G}_{TP} = RT \int_1^X d \ln X,$$

$$\overline{G}_\ell - \overline{G}_\ell^0 = RT \ln X - RT \ln 1,$$

$$\overline{G}_\ell = \overline{G}_\ell^0 + RT \ln X. \tag{49}$$

Referring as it does to the pure solvent, $\overline{G}_\ell^0$ is a constant for all ideal solutions in the given solvent at the given temperature and under the standard pressure.[37]

Defining the ideal solution

To perform a thermodynamic analysis of the ideal gas, we had first to define it—as we did with the equation $PV = nRT$. To perform a thermodynamic analysis of the ideal solution, we have also first to define it with an equation. Now in deriving equation (49) we have, in effect, used

Raoult's law as our definition of the ideal solution to which that equation will apply. But there is an alternative: we may prefer to define the ideal solution directly in terms of equation (49). Instead of deriving (49) from the definition provided by Raoult's law, we would then derive Raoult's law from the definition provided by equation (49). The two alternative definitions differ[38] only by the (generally trivial) margin of inadequacy of the double approximation used, on p. 111, in our derivation of equation (d). Nevertheless, we gain some important advantages by defining the ideal solution in terms of equation (49), and this is the definition we shall henceforth adopt.

One may properly object that this definition is so abstract that, unlike Raoult's law, it offers us no clean-cut empirical criterion for recognition of what will pass as an ideal solution. However, problem 48 invites a demonstration that, under conditions of constant temperature and pressure, *no* change of volume and *no* transfer of heat can attend the mixing of components that meet the standard of ideality set by equation (49). For recognition of the ideal solution so defined, $\Delta H_{\text{mix}} = 0$ and $\Delta V_{\text{mix}} = 0$ thus represent empirical criteria even more readily accessible than that presented by Raoult's law. A further advantage of the definition founded on equation (49) is that it easily accommodates materials too involatile to be meaningfully discussed in terms of Raoult's law. Last, but not least important, definition in terms of equation (49) is rendered highly attractive by the facility with which we can derive from it all the most significant characteristics of ideal solutions.

As an example, let us seek the relation between the molar enthalpy ($\overline{H}_\ell^0$) of a pure liquid solvent and its molar enthalpy ($\overline{H}_\ell$) at the same temperature and pressure in a solution where its behavior conforms to equation (49). The difference $\Delta\overline{H}_\ell$ we define by writing: $\Delta\overline{H}_\ell \equiv \overline{H}_\ell - \overline{H}_\ell^0$. For the corresponding difference $\Delta\overline{G}_\ell$, equation (49) yields

$$\Delta\overline{G}_\ell \equiv \overline{G}_\ell - \overline{G}_\ell^0 = RT \ln X.$$

Now $\Delta\overline{H}_\ell$ and $\Delta\overline{G}_\ell$ are directly linked by the Gibbs-Helmholtz equation derived on p. 105:

$$\frac{-\Delta\overline{H}_\ell}{T^2} = \frac{d}{dT}\left[\frac{\Delta\overline{G}_\ell}{T}\right]_P.$$

Substituting the expression for $\Delta\overline{G}_\ell$, we at once find

$$+\Delta H_\ell = -T^2 \frac{d}{dT}\left[\frac{\Delta\overline{G}_\ell}{T}\right]_P = -T^2 \frac{d}{dT}[R \ln X]_P = 0,$$

since neither R nor X are functions of T. Thus the molar enthalpy ($\overline{H}_\ell^0$) of the pure solvent is shown to be equal to its molar enthalpy ($\overline{H}_\ell$) in an ideal solution at the same temperature and pressure. This result will be of some importance in our consideration of the colligative properties, to which we now turn.

Boiling-point elevation

Consider a dilute solution of an involatile solute in a volatile solvent that conforms to equation (49). At its boiling temperature the solution stands in equilibrium with solvent vapor at 1 atm pressure. In this condition

$$\overline{G}_l = \overline{G}_g,$$

where $\overline{G}_l$ and $\overline{G}_g$ are the molar free energies of the solvent in the liquid and gas phases respectively. If an infinitesimal change is now made in the mole fraction of the solvent, the solution will again stand in equilibrium with solvent vapor at 1 atm pressure only after some infinitesimal change in the boiling temperature. Then

$$\overline{G}_l + d\overline{G}_l = \overline{G}_g + d\overline{G}_g$$

so that, as before,

$$d\overline{G}_l = d\overline{G}_g.$$

We seek now to evaluate $d\overline{G}_l$ and $d\overline{G}_g$.

The pure vapor phase remains constant in composition and pressure, but changes in temperature. On the strength of equation (43) we write

$$d\overline{G}_g = -\overline{S}_g^0\, dT,$$

where $\overline{S}_g^0$ symbolizes the molar entropy of the solvent vapor under a pressure of 1 atm.

In the liquid phase the solvent remains under a pressure of 1 atm, but changes in both temperature and concentration. As before, the total change ($d\overline{G}_l$) can be resolved into the *sum* of a change ($d\overline{G}_{TP}$) due to a shift of concentration at constant temperature *plus* a change ($d\overline{G}_{XP}$) due to a shift of temperature at constant composition. Thus:

$$d\overline{G}_l = d\overline{G}_{TP} + d\overline{G}_{XP} = RT\, d\ln X - \overline{S}_l\, dT,$$

where $\overline{S}_l$ is the molar entropy of the liquid solvent in the solution where its mole fraction is X. Substituting in our equilibrium criterion, $d\overline{G}_l = d\overline{G}_g$, we obtain:

$$RT\, d\ln X - \overline{S}_l\, dT = -\overline{S}_g^0\, dT,$$

$$d\ln X = -\frac{\overline{S}_g^0 - \overline{S}_l}{RT}\, dT.$$

Now $(\overline{S}_g^0 - \overline{S}_l)$ represents the entropy change accompanying the transfer of a mole of solvent from the solution to the vapor phase with which it stands in equilibrium. For such a transfer equation (26) indicates

$\Delta S = \Delta H/T$, so that

$$\overline{S}_{\mathrm{g}}^{0} - \overline{S}_{\ell} = \frac{\overline{H}_{\mathrm{g}}^{0} - \overline{H}_{\ell}}{T},$$

where $\overline{H}_{\mathrm{g}}^{0}$ is the molar enthalpy of the pure solvent vapor, and $\overline{H}_{\ell}$ is the molar enthalpy of the liquid solvent in the ideal solution. But for such a solution we earlier showed that $\overline{H}_{\ell} = \overline{H}_{\ell}^{0}$, where $\overline{H}_{\ell}^{0}$ symbolizes the molar enthalpy of the pure liquid solvent. Hence

$$\overline{S}_{\mathrm{g}}^{0} - \overline{S}_{\ell} = \frac{\overline{H}_{\mathrm{g}}^{0} - \overline{H}_{\ell}^{0}}{T} = \frac{\Delta\overline{H}_{\mathrm{vap}}^{0}}{T},$$

where $\Delta\overline{H}_{\mathrm{vap}}^{0}$ represents the molar enthalpy of vaporization of the pure solvent. Substituting in the last equation of the preceding paragraph, we find

$$d \ln X = -\frac{\Delta\overline{H}_{\mathrm{vap}}^{0}}{RT^{2}}\, dT.$$

Integrating, we take as the upper limits the temperature (T) at which a solution containing some mole fraction (X) of solvent boils under a pressure of 1 atm. As the lower limits we take the temperature (T_B) at which, with mole fraction $X = 1$, the pure solvent boils under a pressure of 1 atm. Approximating only by treating $\Delta\overline{H}_{\mathrm{vap}}^{0}$ as constant over the short temperature range concerned, we write

$$\int_{1}^{X} d \ln X = \frac{\Delta\overline{H}_{\mathrm{vap}}^{0}}{R} \int_{T_B}^{T} -\frac{dT}{T^{2}},$$

$$\ln X = \frac{\Delta\overline{H}_{\mathrm{vap}}^{0}}{R}\left(\frac{1}{T} - \frac{1}{T_B}\right), \qquad (50)$$

$$-\ln X = \frac{\Delta\overline{H}_{\mathrm{vap}}^{0}}{R}\, \frac{(T - T_B)}{TT_B}.$$

With $X < 1$, $-\ln X$ is a positive quantity. And on the right, with all other terms necessarily positive, the difference $(T - T_B)$ must also be a positive quantity. That is, the boiling point of the solution is higher than that of the pure solvent, by a margin which the last equation relates to the mole fraction of solvent in the solution.

Boiling-point elevations are generally measured in very dilute solutions, and it is in such solutions alone that solvents generally conform reasonably well to the standard of ideality set by equation (49). With highly dilute solutions, a number of simplifying assumptions become possible. For example, since in these solutions T will not differ significantly from T_B, we can make the approximation $TT_B \cong T_B^2$. Denoting by δT_B the difference $(T - T_B)$, we now rewrite the last equation as

$$-\ln(1 - X_u) = \frac{\Delta \overline{H}^0_{\text{vap}}}{RT_B^2}\,\delta T_B.$$

Here the mole fraction of the solvent is represented by the difference $(1 - X_u)$, where X_u symbolizes the mole fraction of the solute. But in a very dilute solution $X_u \ll 1$, and a purely mathematical theorem assures us that in these circumstances,

$$-\ln(1 - X_u) = X_u + X_u^2/2 + X_u^3/3 + \cdots \cong X_u.$$

Substituting in the last preceding equation, we thus obtain

$$X_u = \frac{\Delta \overline{H}^0_{\text{vap}}}{RT_B^2}\,\delta T_B.$$

Invoking for the last time the diluteness of the solution, we write

$$X_u = \frac{\text{moles solute}}{\text{moles solute} + \text{moles solvent}} \cong \frac{\text{moles solute}}{\text{moles solvent}}$$

$$= \frac{\text{moles solute per 1000 gm solvent}}{1000/M_{\text{solvent}}} = \frac{mM_{\text{solvent}}}{1000}.$$

Here m represents the molality of the solute, and M_{solvent} the molecular weight of the solvent. Substitution now gives

$$\frac{mM_{\text{solvent}}}{1000} = \frac{\Delta \overline{H}^0_{\text{vap}}\,\delta T_B}{RT_B^2},$$

$$\delta T_B = \left[\frac{RT_B^2 M_{\text{solvent}}}{1000\,\Delta \overline{H}^0_{\text{vap}}}\right] m. \qquad (51)$$

For any given solvent, all the terms in the bracketed expression are constants, and can be lumped together in a single constant (K_B) characteristic of the solvent in question. Whatever the involatile solute concerned, all ideal dilute solutions in the given solvent will thus conform to the remarkably simple equation: $\delta T_B = K_B m$. And despite the many approximations involved in our derivation of equation (51), the bracketed constellation of constants yields calculated values for K_B that are handsomely confirmed by experimental measurements—as shown in Table 5.

▶ *Example 5*

Under a pressure of 1 atm, carbon disulfide boils at 46.29°C with the absorption of 84.1 cal per gm CS_2 vaporized. Under the same pressure,

TABLE 5

Solvent	Boiling-point at 1 atm, °K	$\Delta\overline{H}^0_{vap}$ kcal/mole	K_B	
			calc.	obs.
Water	373.2	9.72	0.513	0.51
Acetone	329.3	7.28	1.72	1.72
Carbon tetrachloride	349.9	7.17	5.22	4.9
Chloroform	334.4	7.02	3.78	3.9
Ethyl alcohol	351.7	9.22	1.23	1.20
Methyl alcohol	337.9	8.43	0.86	0.84
Diethyl ether	307.6	6.61	2.11	2.16
Benzene	353.3	7.35	2.63	2.6

a solution of 0.3 gm of sulfur in 10 gm of carbon disulfide boils at 46.57°C. Determine (a) the boiling-point-elevation constant (K_B) of carbon disulfide; and (b) the molecular state of the dissolved sulfur (atomic weight = 32).

Solution. (a) In terms of the molecular weight of carbon disulfide ($M_{solvent}$), we can write the molar heat of vaporization as

$$\Delta\overline{H}^0_{vap} = 84.1 \times M_{solvent}.$$

The bracketed term in equation (51) yields

$$K_B = \frac{RT_B^2 M_{solvent}}{1000\ \Delta\overline{H}^0_{vap}} = \frac{RT_B^2}{84100} = \frac{(1.99)(319.44)^2}{84100} = 2.41.$$

(b) If we could determine the molecular weight (M) of the dissolved sulfur, we could infer something about its condition in solution. With 30 grams of sulfur per 1000 grams of solvent, the molality (m) of the dissolved sulfur can be expressed as $m = 30/M$. The boiling-point elevation of this solution is given as $\delta T_B = 46.57 - 46.29 = 0.28°$. Substituting in the equation $\delta T_B = K_B m$, we find

$$0.28 = 2.41\frac{30}{M} \qquad \text{or} \qquad M = 258.$$

Only the first two figures are significant, but they supply ample basis for the assignment to the dissolved sulfur of the molecular formula S_8, i.e., $8 \times 32 = 256$. ◀

Freezing-point depression

At its freezing point, a solution stands in equilibrium with pure solid solvent under a pressure of 1 atm. With the pressure held constant, an infinitesimal change in the concentration of the solution produces a new equilibrium state characterized by an infinitesimally different freezing temperature. For the difference between the two equilibrium states, we can write as before

$$d\overline{G}_\ell = d\overline{G}_s,$$

where $\overline{G}_\ell$ and $\overline{G}_s$ symbolize the molar free energies of the solvent in the liquid and solid phases respectively.

In the pure solid phase the only change is one of temperature, and equation (43) yields

$$d\overline{G}_s = -\overline{S}_s^0\, dT,$$

where $\overline{S}_s^0$ symbolizes the molar entropy of the solid solvent under a pressure of 1 atm. In the liquid phase the solvent remains under 1 atm pressure, but changes in both temperature and concentration. If in this solution the solvent conforms to equation (49), we can easily represent $d\overline{G}_\ell$ as the sum of two terms:

$$d\overline{G}_\ell = d\overline{G}_{TP} + d\overline{G}_{XP} = RT\, d\ln X - \overline{S}_\ell\, dT,$$

where $\overline{S}_\ell$ represents the molar entropy of the solvent in the solution where its mole fraction is X. Substituting in the last equation of the preceding paragraph, we find

$$RT\, d\ln X - \overline{S}_\ell\, dT = -\overline{S}_s^0\, dT$$

$$d\ln X = \frac{\overline{S}_\ell - \overline{S}_s^0}{RT}\, dT.$$

Now $(\overline{S}_\ell - \overline{S}_s^0)$ represents the entropy change accompanying the transfer of a mole of solvent from the pure frozen solid to the solution with which it stands in equilibrium. Here too equation (26) can be applied to express $(\overline{S}_\ell - \overline{S}_s^0)$ in terms of $(\overline{H}_\ell - \overline{H}_s^0)$, where $\overline{H}_s^0$ is the molar enthalpy of the pure solid solvent. And the molar enthalpy $(\overline{H}_\ell)$ of the solvent in solution, we can again equate with the molar enthalpy $(\overline{H}_\ell^0)$ of the pure liquid solvent. Hence:

$$\overline{S}_\ell - \overline{S}_s^0 = \frac{\overline{H}_\ell - \overline{H}_s^0}{T} = \frac{\overline{H}_\ell^0 - \overline{H}_s^0}{T} = \frac{\Delta\overline{H}_{\text{fus}}^0}{T},$$

where $\Delta\overline{H}_{\text{us}}^0$ is the molar enthalpy of fusion of the pure solid. By substituting in the last equation of the preceding paragraph, we thus attain:

$$d\ln X = \frac{\Delta\overline{H}_{\text{fus}}^0}{RT^2}\, dT.$$

Integrating, we take as the upper limits the temperature (T) at which a solution containing some mole fraction (X) of solvent freezes under a pressure of 1 atm. As the lower limits we take the temperature (T_F) at

TABLE 6

Solvent	Freezing-point, °K	$\Delta\overline{H}^0_{\text{fus}}$, kcal/mole	K_F	
			calc.	obs.
Water	273.2	1.436	1.86	1.86
Benzene	278.7	2.378	5.07	5.12
Naphthalene	353.4	4.565	6.97	6.9
Carbon tetrachloride	250.3	0.60	32	30
Chloroform	209.7	2.2	4.7	4.9
Ethylene dibromide	283.1	2.62	11.4	12.5
Acetic acid	289.8	2.80	3.6	3.9
Phenol	314.2	2.70	6.8	7.3

which (with mole fraction $X = 1$) the pure solvent freezes under a pressure of 1 atm. Approximating only by treating $\Delta\overline{H}^0_{\text{fus}}$ as constant over the short temperature range concerned, we obtain by integration:

$$\ln X = -\frac{\Delta\overline{H}^0_{\text{fus}}}{R}\left(\frac{1}{T} - \frac{1}{T_F}\right), \tag{52}$$

$$-\ln X = \frac{\Delta\overline{H}^0_{\text{fus}}}{R}\,\frac{(T_F - T)}{TT_F}.$$

Again $(-\ln X)$ must be a positive quantity. And on the right all terms save $(T_F - T)$ must also be positive. Hence the difference $(T_F - T)$ is also a positive quantity; i.e., the solution freezes at a temperature lower than the freezing point of the pure solvent.

For the very dilute solutions in which freezing-point depressions are generally measured, a series of approximations identical with those of the preceding section now yields

$$\delta T_F = \left[\frac{RT_F^2 M_{\text{solvent}}}{1000\,\Delta\overline{H}^0_{\text{fus}}}\right] m = K_F m, \tag{53}$$

where δT_F symbolizes the difference $(T_F - T)$. As shown in Table 6, values for K_F calculated from this equation are handsomely confirmed by experimental measurements.

Solubility: A colligative property?

From freezing-point depressions to solubilities is but a short step. At the left in Fig. 34 (after Everett), pure crystalline A is shown in equilibrium with a liquid phase containing both A and B. Observing such a system, we find that the equilibrium temperature and the composition of the

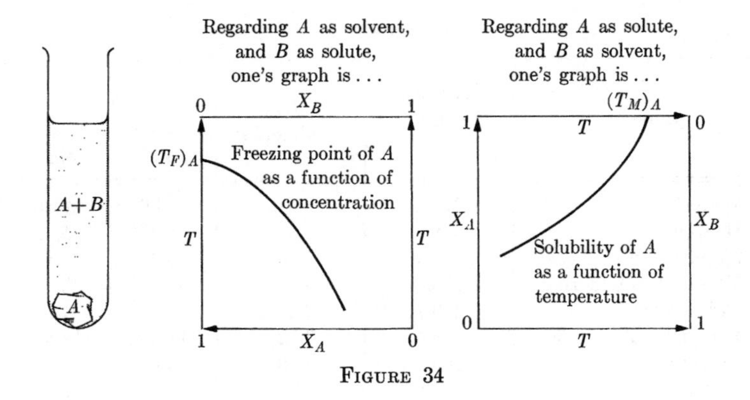

FIGURE 34

liquid are mutually dependent. Now according to whether we regard A or B as the solvent, we will be led to display the very same observations in very different ways. *Either* we plot the freezing point of A as a function of its concentration in the solution, as in the middle panel. *Or* we plot the solubility of A as a function of temperature, as at the right. The two plots differ only by a 90° rotation. What distinguishes phenomena of freezing-point depression and solubility thus comes down to a difference of point of view and terminology. We may then hope to construe solubility data in terms of a relation derived for freezing-point depressions.

The multiple assumptions of great dilution used in deriving equation (53) render that relation inapplicable to solubility phenomena that involve more concentrated solutions. But no such difficulty impedes application of the parent equation (52) to solubility data. How shall we proceed? What we formerly took to be the solvent (i.e., the material that separates out as a pure solid phase) will now be called the solute (i.e., the material that dissolves to form a saturated solution). And correspondingly, what we formerly called the freezing point (T_F) of the pure solvent will now be called the melting point (T_M) of the pure solute. These two small changes we make explicit when we rewrite equation (52) as

$$\ln X_u = -\frac{\Delta \overline{H}^0_{\text{fus}}}{R}\left(\frac{1}{T} - \frac{1}{T_M}\right). \tag{e}$$

Permitting the calculation of solubility at all temperatures up to the solute's melting point, this equation surprisingly suggests that—in terms of mole fraction—the solubility of a given substance at a given temperature should be the *same* in all solvents with which it forms ideal solutions. And problem 51 invites a demonstration that, the lower the melting point and the lower the heat of fusion of the substance concerned, the greater will be its ideal solubility at any given temperature. The powerful solvation

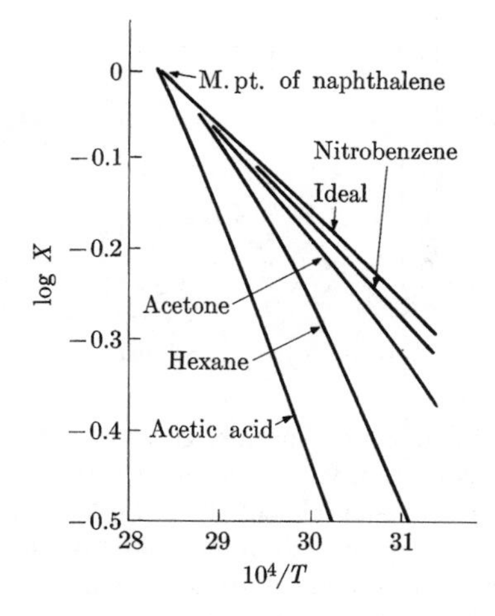

FIG. 35. Solubility of naphthalene in various solvents, as a function of temperature.

effects that accompany the solution of electrolytes disqualify these solutions as ideal. But, illustrative of a more favorable instance, Fig. 35 shows the extent to which the solubility of naphthalene in various solvents is approximated by equation (e). Note that in several cases the plot of $\log X$ *vs* $1/T$ very well approximates the predicted straight line.[39] Observe also the predictable convergence of *all* curves when, in the limit of pure naphthalene, $T \rightarrow T_M$ as $X_u \rightarrow 1$.

Recognizing the relation between *freezing-point depressions* and the solubility of *solids*, we are led to seek an analogous relation between *boiling-point elevations* and the solubility of *gases*. Review the chain of argument by which we passed, from equation (52) and freezing-point depressions, to equation (e) and the solubility of solids. Precisely the same line of argument will carry us, from equation (50) and boiling-point elevations, to the following expression for the variation with temperature of the solubility of a gas maintained at *constant pressure:*

$$\ln X_u = \frac{\Delta \overline{H}^0_{\text{vap}}}{R}\left(\frac{1}{T} - \frac{1}{T_B}\right).$$

If one atmosphere is to be the pressure maintained, T_B will be the normal boiling point of the pure liquefied gas. But if some other constant pressure were to be used, we would have only to substitute for T_B whatever *is* the boiling point of the pure liquefied gas under *that* pressure. At a given temperature and pressure, the last equation indicates that the solubility of a given gas should bring it to the *same* mole fraction in all solvents with

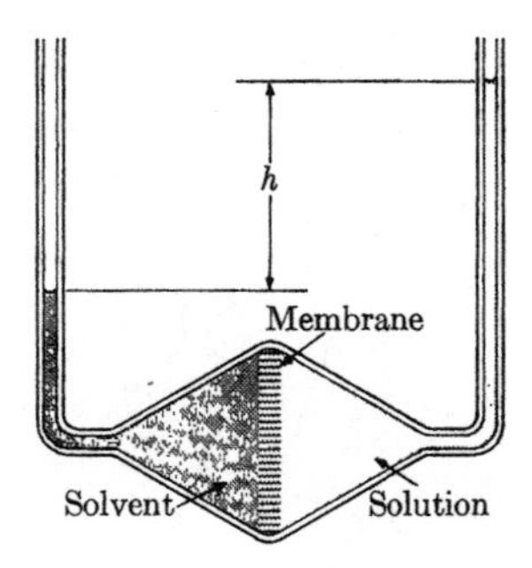

FIGURE 36

which it forms ideal solutions. In this, gaseous and solid solutes behave alike; but there is also an important difference that reflects the presence in equation (e) of a minus sign absent from the last equation. And, to be sure, while solids generally *increase* in solubility with rising temperature, the last equation permits us to calculate by how much the ideal solubilities of gases *decrease* with rising temperature.

Osmotic pressure

Let a pure solvent be separated from one of its solutions, standing at the same temperature and pressure, by a membrane selectively permeable to the solvent only. In these circumstances a net transport of solvent, across the membrane into the solution, acts to produce a pressure differential like that corresponding to the hydrostatic head (h) shown in Fig. 36. When this pressure differential has attained some particular magnitude, the solution stands at equilibrium with the pure solvent, of which there is then no longer any net transfer. This particular pressure differential we call the osmotic pressure (π), and we seek now to express it as a function of the mole fraction (X) of solvent in the solution concerned.[40]

When solvent and solution stand at equilibrium with each other,

$$\overline{G}_l = \overline{G}_l^0.$$

As before, $\overline{G}_l$ here represents the molar free energy of the solvent in the solution, while $\overline{G}_l^0$ symbolizes the molar free energy of the pure solvent under a pressure of 1 atm at the prevailing temperature. If the concentration of the solution is slightly altered, a new condition of equilibrium is reached when, under a slightly altered pressure, the solution again stands in equilibrium with the *same* pure solvent under the *same* pressure of 1 atm at the *same* prevailing temperature. For this new condition of equilibrium we write

$$\overline{G}_l + d\overline{G}_l = \overline{G}_l^0,$$

whence it follows that

$$d\overline{G}_l = 0.$$

That is, the effect of the pressure change so counteracts the effect of the concentration change that there is *no* net change in the molar free energy of the solvent in the solution. For an ideal solution we can readily formulate the mutual cancellation of the two changes, by writing

$$d\overline{G}_\ell = d\overline{G}_{XT} + d\overline{G}_{TP} = \overline{V}_\ell \, dP + RT \ln X = 0,$$

where $\overline{V}_\ell$ symbolizes the molar volume of the solvent in the solution.

If we assume that the solvent is essentially incompressible over a short range of pressure, we can treat $\overline{V}_\ell$ as a constant. Integration of the last equation then presents no problems.

$$\overline{V}_\ell \int_1^{1+\pi} dP + RT \int_1^X \ln X = 0.$$

As upper limits we have taken the concentration (X) at which a solution under pressure $(1 + \pi)$ atm will stand in equilibrium with pure solvent under a pressure of 1 atm at the same temperature. As lower limits we take the concentration $X = 1$ at which the solution compartment would be occupied by pure solvent which, under a pressure of 1 atm, would stand in equilibrium with the pure solvent in the reference chamber. And so

$$\pi \overline{V}_\ell + RT \ln X = 0. \tag{54}$$

For the very dilute solutions in which osmotic pressures are generally measured, a familiar series of approximations now yields

$$\pi = \frac{RT}{\overline{V}_\ell}(-\ln X) \cong \frac{RT}{\overline{V}_\ell} X_u \cong \frac{RT}{\overline{V}_\ell} \cdot \frac{\text{moles solute present}}{\text{moles solvent present}}.$$

But in a dilute solution the moles of solvent times the molar volume thereof will not differ significantly from the *total* volume (V) of the solution. Symbolizing by n the number of moles of solute in that volume, we thus arrive at van't Hoff's osmotic-pressure law:

$$\pi V = nRT. \tag{55}$$

Like the ideal-gas law it so resembles, this relation is at its best as a limiting law for the extreme of infinite dilution. If what we seek is the molecular weight (M) of the solute, we can facilitate the extrapolation to infinite dilution by rewriting the last equation—first in terms of the weight (w) of solute present, and then in terms of its concentration (c) in grams per liter. Thus

$$\pi V = \frac{w}{M} RT,$$

$$M = \frac{1}{\pi}\frac{w}{V} RT = \frac{c}{\pi} RT = \frac{RT}{\pi/c}.$$

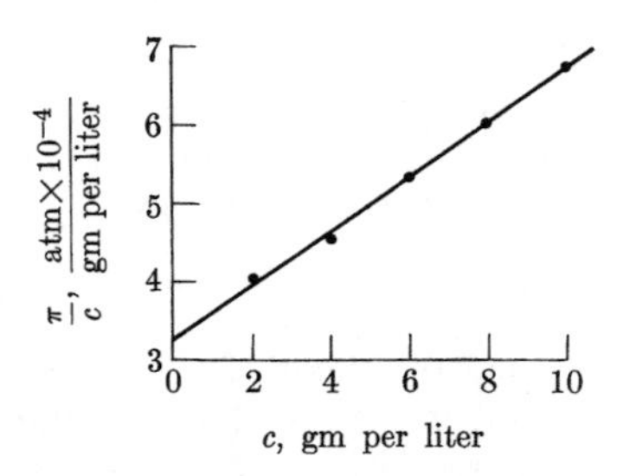

FIGURE 37

The ratio π/c may vary markedly with c, as indicated in Fig. 37 for solutions of a polyvinylchloride. A respectably accurate value for M is then attainable only by using the limiting value of π/c established by extrapolation to $c = 0$.

▶ *Example 6*

The measurements plotted in Fig. 37 were made at 40°C on solutions of polyvinylchloride in cyclohexanone, and the extrapolation yields 3.3×10^{-4} atm/gm per liter as the limiting value of π/c. Calculate the molecular weight of the polyvinylchloride.

Solution. With π expressed in atm, and c a concentration per liter, the appropriate value for R will be 0.0821 lit-atm/mole-°K. Then:

$$M = \frac{(0.0821)(313)}{3.3 \times 10^{-4}} = 7.8 \times 10^4.$$

From a much more careful process of curve fitting, a value of 79000 is obtained by the authors* whose data we have used in preparing Fig. 37.

EQUILIBRIUM STATE AND EQUILIBRIUM CONSTANT

A reaction mixture of ideal gases offers a peculiarly simple context for this next approach to the phenomenon of chemical equilibrium. For a mole of some *one* pure ideal gas at a fixed temperature, we obtained from equation (43)

$$d\overline{G} = \overline{V}\,dP = \frac{RT}{P}\,dP = RT\,d\ln P.$$

Integration at once yields the relation of the molar free energy ($\overline{G}$) of the gas at any pressure (P) to the standard molar free energy ($\overline{G}^0$) of the gas

---

* Bonnar, Dimbat, and Stross, *Number Average Molecular Weights* (Interscience, 1958), p. 258.

at unit pressure ($P = 1$):

$$\overline{G} - \overline{G}^0 = RT \ln P/1.$$

We have written $P/1$ to call attention to an important matter of units. The units of the $\overline{G}$'s—usually kcal/mole or joules/mole—are of course determined by the units of R in the RT-term on the right; and the logarithmic term involves only a properly dimensionless *ratio* of pressures. But, whatever the units of $\overline{G}^0$, it refers to the molar free energy of the pure gas in a defined standard state, under some one particular *unit pressure* at the given temperature. Possible unit pressures are 1 cm Hg, 1 atm, etc., and to each alternative unit-pressure standard state must correspond a different value of $\overline{G}^0$. This linkage is less obvious when the last equation is rewritten, as it usually is, in the form:

$$\overline{G} - \overline{G}^0 = RT \ln P. \tag{56}$$

Appearances notwithstanding, the logarithmic term still involves a pressure ratio, of which the denominator (1) has simply been suppressed. And since $\overline{G}^0$ refers to a standard state defined with *that* unit pressure, we must take care always to express P in terms of that pressure unit. In this text we always define the standard state as at 1 atm pressure, and for us atm will then always be the appropriate units for P in the last equation.

To gain an expression for the molar free energy of a component in an ideal solution, in an earlier section we used Raoult's law as a provisional definition in deriving the equation $\overline{G}_l - \overline{G}_l^0 = RT \ln X$. But then we adopted that equation as our final definition of the ideal solution and, at this point, we revise our definition of the ideal gas in the very same way. Heretofore the empirical limiting law, $P\overline{V} = RT$, has served as our definition—as above, where $\overline{V}$ is replaced by RT/P in deriving the equation $\overline{G} - \overline{G}^0 = RT \ln P$. But we now adopt this equation as our final definition of what we mean by an ideal gas; and problem 56 invites the demonstration that, by a simple inversion of the argument, the relation $P\overline{V} = RT$ can easily be derived from the defining equation. Proceeding further in the same direction, we now define the *ideal mixture of ideal gases* as one in which the molar free energy of the ith component conforms to the equation:

$$\overline{G}_i - \overline{G}_i^0 = RT \ln P_i. \tag{57}$$

Here P_i symbolizes the partial pressure of the ith component in the mixture. And the standard free energy $\overline{G}_i^0$ again represents the molar free energy of the corresponding pure gas under the standard pressure of 1 atm at the same temperature.

The near identity of equations (56) and (57) is rendered eminently plausible by our conception of ideal gases as noninteracting point masses.

And at reasonably low pressures, actual mixtures of real gases are found to agree quite closely with the empirical consequences that follow from the equation we have used to define the ideal gas mixture.[41] Indeed, a mixture of real gases that conforms well to equation (57) is encountered far more often in practice than is a real solution that well approximates the ideal solution defined by equation (49).

When, at a partial pressure P_i, the ith component of an ideal gas mixture is present therein to the extent of n_i moles, equation (57) assumes the form:

$$G_i = n_i\overline{G}_i = n_i\overline{G}_i^0 + n_iRT \ln P_i.$$

And now we are in a position to discuss chemical equilibrium as it occurs in ideal gas mixtures. As our type reaction we take:

$$aA + bB = cC + dD.$$

What change of free energy (ΔG) occurs when a moles of substance A at partial pressure P'_A react with b moles of B at pressure P'_B, to form c moles of C at pressure P'_C and d moles of D at pressure P'_D? We can as usual write

$$\Delta G = c\,\overline{G}_C + d\,\overline{G}_D - (a\,\overline{G}_A + b\,\overline{G}_B).$$

Our definition of the ideal gas mixture now permits us to express ΔG in terms of the quoted partial pressures.

$$\begin{aligned}\Delta G &= c\overline{G}_C^0 + cRT \ln P'_C + d\overline{G}_D^0 + dRT \ln P'_D - a\overline{G}_A^0 - aRT \ln P'_A - b\overline{G}_B^0 - bRT \ln P'_B \\ &= (c\overline{G}_C^0 + d\overline{G}_D^0 - a\overline{G}_A^0 - b\overline{G}_B^0) + RT(c \ln P'_C + d \ln P'_D - a \ln P'_A - b \ln P'_B).\end{aligned}$$

Symbolizing by ΔG^0 the indicated difference of standard free energies, we find for the reaction as written

$$\Delta G = \Delta G^0 + RT \ln \frac{(P'_C)^c(P'_D)^d}{(P'_A)^a(P'_B)^b}.$$

We have thus attained an expression for the change of free energy that accompanies the conversion of reactants at certain partial pressures into products at certain partial pressures. Were it the case that

$$P'_A = P'_B = P'_C = P'_D = 1 \text{ atm},$$

the logarithmic term would reduce to zero, and then $\Delta G = \Delta G^0$. This would be entirely appropriate, since the reaction for which we calculate ΔG would now convert reactants at the *standard* pressure into products at the *standard* pressure. But the really central importance of the above equation emerges only when we consider its bearing on the *equilibrium* state of the gas mixture. Let P_A, P_B, P_C, and P_D symbolize the partial pressures of the respective species in some particular mixture in which they stand at equilibrium. An equilibrium state we identified as a state of minimum free energy for which our criterion is still $\Delta G = 0$. That is, the conversion of reactants at equilibrium partial pressures, into products at equilibrium partial pressures, or *vice versa,* would be a change for which $\Delta G = 0$. Therefore

$$0 = \Delta G = \Delta G^0 + RT \ln \frac{(P_C)^c(P_D)^d}{(P_A)^a(P_B)^b}$$

and

$$\ln \frac{(P_C)^c(P_D)^d}{(P_A)^a(P_B)^b} = -\frac{\Delta G^0}{RT}.$$

A remarkable conclusion now lies before us. Unlike ΔG, ΔG^0 is wholly independent of the initial partial pressures. For ΔG^0 refers to a defined difference of *standard* free energies measured, at the temperature in question, with each pure gas under the *standard* pressure of 1 atm. Thus ΔG^0 is a *constant* characteristic of the given reaction at the given temperature. For that temperature, then, the entire right side of the last equation must be a *constant.* For the given temperature, the left side of the equation must then also be a *constant.* Now, depending on how the reaction mixture was made up, equilibrium can be reached with very different sets of individual partial pressures for the various reactants and products. But what we have established is that, at a given temperature, the indicated function of equilibrium partial pressures is a *constant* (hereafter symbolized as K_p). From a purely thermodynamic analysis—invoking neither empirical findings nor hypotheses about the mechanisms of chemical reactions—we have thus deduced the existence of what we call the "equilibrium constant."

We can easily extend this major conclusion to reactions occurring in ideal solutions. When equation (49), $\overline{G}_\ell = \overline{G}_\ell^0 + RT \ln X$, is applicable to all components that figure in the type reaction $aA + bB = cC + dD$, the same line of argument will yield the very similar result:

$$\Delta G_\ell = \Delta G_\ell^0 + RT \ln \frac{(X'_C)^c(X'_D)^d}{(X'_A)^a(X'_B)^b}.$$

Here ΔG_ℓ represents the change in free energy accompanying the conversion of reactants at mole fractions X'_A and X'_B to products at mole fractions X'_C and X'_D in the solution. Symbolizing by X_A, X_B, X_C, and X_D the *equilibrium* mole fractions in any particular reaction mixture, for that equilibrium condition we can write

$$0 = \Delta G_\ell = \Delta G_\ell^0 + RT \ln \frac{(X_C)^c(X_D)^d}{(X_A)^a(X_B)^b}$$

and

$$\ln \frac{(X_C)^c(X_D)^d}{(X_A)^a(X_B)^b} = \frac{-\Delta G_\ell^0}{RT}.$$

What appears on the right is a difference of *standard* free energies. And so, at a given temperature, the right side of the equation is a *constant*. At a given temperature the left side must then also be a *constant*. Thus the indicated function of equilibrium mole fractions again represents an equilibrium constant (symbolized K_X).*

Further to extend this concept, to *non*ideal mixtures of liquids or gases, we have only to introduce the idea of "activity." In a given medium at a given temperature, the activity of a substance is a single-valued monotonic function of its concentration. Sometimes the activity is little more than a "corrected concentration," obtained by multiplying each actual concentration by some suitable numerical coefficient. But, due to molecular interactions wholly ignored in our dealings with ideal gases, the activity is often a very complicated function of concentration. This function we may seek to establish either through experimental measurements or from (nonthermodynamic) theoretical analysis. However it is determined, the activity (a_j) of the jth component in a nonideal mixture or solution makes possible a very familiar form of expression for the molar free energy $\overline{G}_j$:

$$\overline{G}_j - \overline{G}_j^{\Delta} = RT \ln a_j.$$

Here $\overline{G}_j^{\Delta}$ represents the molar free energy of the component in a strictly hypothetical but well defined standard state of unit activity. And setting out from this last equation, as formerly we started from equations (49) or (57), we very soon arrive at the equation: $\ln K_a = -\Delta G^{\Delta}/RT$. With the difference of standard free energies (ΔG^{Δ}) a constant for a given reaction in a given medium at a given temperature, we conclude that the function of equilibrium activities we have already symbolized as K_a is again an equilibrium *constant*. The concept of equilibrium constant is thus established with full generality.

To provide a capsule summary of our findings, we now drop all distinction between the various kinds of standard states. What we have established is that the free energy change (ΔG) accompanying a chemical reaction is always given by an equation of the form:

$$\Delta G = \Delta G^0 + RT \ln \mathcal{Q}. \tag{58}$$

Here ΔG^0 is a defined difference of standard free energies. And the reaction quotient ($\mathcal{Q}$) is a function of partial pressures, or mole fractions, or activities, etc. which has the *form* of an equilibrium constant. For the reaction in which reactants at equilibrium concentrations would be con-

---

* Equilibrium constants expressed in the more familiar terms of molarity and molality are easily attainable by either of the approaches indicated in problems 59 and 60.

verted to products at equilibrium concentrations, $\Delta G = 0$ and $\mathcal{Q}$ takes on the *value* of the equilibrium constant (K). And then

$$\ln K = -\Delta G^0/RT. \tag{59}$$

From known values of ΔG^0 we can now calculate unknown equilibrium constants. What if $\Delta G^0 = 0$? Equation (59) then requires that $K = 1$. And, to be sure, both these equalities alike imply that equilibrium will reign in a reaction mixture where each reactant and each product is present at unit activity. If on the other hand $\Delta G^0 < 0$, equation (58) indicates that equilibrium will *not* prevail in such a reaction mixture. The equilibrium criterion $\Delta G = 0$ can now be satisfied only after, with consumption of reactants and generation of products, the reaction has proceeded some distance to the right. And of course when $\Delta G^0 < 0$, equation (59) duly yields $K > 1$—which is simply another way of saying that equilibrium lies to the right, and just how far to the right it lies. Finally, if $\Delta G^0 > 0$ equation (59) yields $K < 1$, and both these inequalities signify alike that, as written, the reaction proceeds spontaneously not to the right but to the left. And the larger the margin by which $\Delta G^0 > 0$, the greater must be the extent to which $K < 1$, and the further to the left must lie the position of equilibrium.[42]

▶ *Example 7*

Until the 20th century, utter failure attended every attempt to bring about the reaction

$$\tfrac{1}{2}\,N_2\ (g) + \tfrac{3}{2}\,H_2\ (g) = NH_3\ (g).$$

If failure is due merely to slowness of the reaction, further effort (e.g., the search for a catalyst) may be justifiable—as it would *not* be if failure is due to thermodynamic impossibility reflected in a hopelessly unfavorable equilibrium constant. From the following thermodynamic data, calculate K_p for the above reaction at 298°K.

	ΔH_f^0, kcal/mole	S_{298}^0, cal/mole-°K
N_2 (g)	—	45.77
H_2 (g)	—	31.21
NH_3 (g)	−11.02	45.97

Solution. For the indicated reaction we see at once that

$$\Delta H_{298}^0 = -11.02\ \text{kcal/mole},$$

and to establish the corresponding value of ΔS^0_{298} we have only to write

$$\Delta S^0_{298} = 45.97 - \tfrac{1}{2}(45.77) - \tfrac{3}{2}(31.21) = -23.73 \text{ cal/mole-°K}.$$

Equation (37) now yields for ΔG^0_{298}

$$\begin{aligned}\Delta G^0_{298} &= \Delta H^0_{298} - T\,\Delta S^0_{298}\\ &= -11{,}020 - 298(-23.73) = -3950 \text{ cal/mole}.\end{aligned}$$

This amply indicates the thermodynamic feasibility of the reaction in question, for which the equilibrium constant is now easily calculable as

$$\log K_p = \frac{-\Delta G^0_{298}}{2.30\ RT} = \frac{+3950}{(2.30)(1.99)(298)} = 2.90,$$

$$K_p = \frac{P_{NH_3}}{(P_{N_2})^{1/2}(P_{H_2})^{3/2}} = 7.9 \times 10^{+2}.$$ ◀

Because of the logarithmic form of equation (59), even a crude measurement of K will yield a reasonably accurate figure for ΔG^0. But, for the same reason, a respectably accurate value for K will be obtainable only from a *highly* accurate figure for ΔG^0. Sometimes a figure of the requisite accuracy can most easily be drawn from measurements made on a galvanic cell.

The galvanic cell

Unlike any system so far considered, the galvanic cell is a device capable of supplying net (electrical) work in excess of $P\,dV$ work. For the reaction occurring in the cell, we may then hope to use equation (38) to determine ΔG^0 directly from a measurement of the maximum output of electrical work. Under what conditions might we attain such a measurement, corresponding to effectively reversible operation of the cell? Consider that if the cell is simply shortcircuited, *no* work is recovered—just as when an ideal gas expands into a vacuum. Maximum recovery of $P\,dV$ work demands, we found, that the external pressure shall at no time fall more than infinitesimally short of the internal pressure. And maximum recovery of electrical work similarly demands that, when charge is drawn from a galvanic cell, the voltage opposing the current shall at no time fall more than infinitesimally short of the cell EMF that drives the current. An infinitesimal increase in the opposing voltage then suffices to reverse the direction of the current, which is itself always restricted to infinitesimal magnitudes. Reversible operation of the cell will be approachable only under these exacting conditions which, very fortunately, are almost perfectly realized in use of the Poggendorf potentiometer. This instrument yields a null measurement of the cell's EMF, by opposing to it a precisely measurable external voltage which is adjusted until no detectable current enters or leaves the cell. The adjustment can be made with such delicacy that the residual current need not exceed 10^{-9} ampere, and a mole of reaction in the cell then requires the comparatively "infinite" time of 3×10^6 years. In such a study of the galvanic cell, we actually make our closest approach to realizing in practice the theoretical condition of reversibility.*

As an example of a redox reaction that can be brought about in a galvanic cell, consider

$$\text{Zn (s)} + \text{Cu}^{++} = \text{Cu (s)} + \text{Zn}^{++}.$$

This reaction proceeds whenever current is drawn from the galvanic cell constituted by linking copper and zinc half-cells with a salt bridge. Let a potentiometer be used to determine the difference of potential ($\mathfrak{E}$) delivered by the galvanic cell. For the free-energy change in the cell reaction, equation (38) then yields

$$-\Delta G = w_{\text{net}} = \text{volts} \times \text{coulombs},$$
$$-\Delta G = \mathfrak{E} \times n\mathfrak{F}, \tag{60}$$

where $\mathfrak{F}$ represents the faraday ($\cong$ 96,500 coulombs) and n symbolizes the number of electrons transferred in the equation we have written for the cell reaction. With a pair of half-cells in which all components are present in their standard states and concentrations, the corresponding equation will be

$$-\Delta G^0 = n\mathfrak{F}\mathfrak{E}^0,$$

where $\mathfrak{E}^0$ is the measured potential difference between the *standard* half-cells. In favorable cases this difference is easily determinable with an uncertainty less than ± 0.001 volt. The last equation then permits us to establish, for a reaction in which $n = 1$, a value of ΔG^0 good to ± 23 cal/mole. Since ΔG^0 may run to many tens of *kilo*calories, one readily understands how the potentiometric method can yield a truly excellent value for ΔG^0, and often the very best value accessible.[43]

---

* Joulean heating (= the resistive dissipation of heat that is the electric analog of frictional loss in mechanical systems) becomes negligible, compared with the power recoverable from the cell, only in the limit where the current ($\mathcal{I}$) approaches zero. Symbolizing by $\mathcal{R}$ the internal resistance of the cell, and its voltage by $\mathfrak{E}$, we can express the limiting condition as:

$$\frac{\text{Power loss due to Joulean heating}}{\text{Power output}} = \frac{\mathcal{I}^2\mathcal{R}}{\mathcal{I}\mathfrak{E}} = \frac{\mathcal{I}\mathcal{R}}{\mathfrak{E}} \rightarrow 0 \qquad \text{when } \mathcal{I} \rightarrow 0.$$

When potentiometric results are available, we may substitute from equation (60) in equation (58), and so obtain

$$n\mathfrak{F}\mathfrak{E} = n\mathfrak{F}\mathfrak{E}^0 - RT \ln \mathfrak{Q},$$

$$\mathfrak{E} = \mathfrak{E}^0 - \frac{2.3RT}{n\mathfrak{F}} \log \mathfrak{Q}. \tag{61}$$

At equilibrium $\mathfrak{E}$, like ΔG, must drop to zero. And the reaction quotient $\mathfrak{Q}$ then assumes the value of the equilibrium constant K, which we can now easily calculate from the relation,

$$\log \mathrm{K} = \frac{n\mathfrak{F}}{2.3RT} \mathfrak{E}^0 \underset{\text{At } 298°\text{K}}{=} \frac{n}{0.059} \mathfrak{E}^0.$$

As before, $\mathfrak{E}^0$ here represents the potential *difference* between two *standard* half-cells. The last equation thus teaches us how the equilibrium constants of redox reactions can at once be calculated from readily available tabulations of standard half-cell potentials, $\mathcal{E}^0$.*

Beyond ΔG, potentiometric measurements on galvanic cells can supply also figures for the ΔS and ΔH characteristic of the cell reaction in question. From equation (44) we know that, under constant-pressure conditions, $d(\Delta G) = -\Delta S\, dT$. Hence

$$\Delta S = -\left[\frac{d(\Delta G)}{dT}\right]_P = -\left[\frac{d(-n\mathfrak{F}\mathfrak{E})}{dT}\right]_P = +n\mathfrak{F}\left[\frac{d\mathfrak{E}}{dT}\right]_P,$$

where $[d\mathfrak{E}/dT]_P$ symbolizes the temperature coefficient of the cell potential under constant-pressure conditions. Substituting in equation (37) our expressions for ΔG and ΔS, we now easily obtain the corresponding expression for ΔH.

$$\Delta H = \Delta G + T\,\Delta S = -n\mathfrak{F}\mathfrak{E} + Tn\mathfrak{F}\left[\frac{d\mathfrak{E}}{dT}\right]_P = n\mathfrak{F}\left(T\left[\frac{d\mathfrak{E}}{dT}\right]_P - \mathfrak{E}\right).$$

* The familiar Nernst equation for half-cell potentials is of course easily derivable from equation (61). Imagine a half-cell in which a atoms, molecules, or ions of some material in its oxidized form (Ox) pass over into b atoms, molecules, or ions of that material in its reduced form (Red), with the transfer of n electrons, according to the reaction a Ox + ne = b Red. Let this half-cell be coupled with a standard hydrogen half-cell in which the reaction is $H^+ + e = \frac{1}{2} H_2$. For this half-cell, by definition, $\mathcal{E} = \mathcal{E}^0 = 0$ when the pressure of gaseous hydrogen is 1 atm and the concentration (better, activity) of hydrogen ion is also 1. Substitution in equation (61) then yields for the first half-cell the following fundamental relation of electrochemistry:

$$\mathcal{E} = \mathcal{E}^0 - \frac{2.3RT}{n\mathfrak{F}} \log \frac{(\mathrm{Red})^b}{(\mathrm{Ox})^a}.$$

With all cell components in their standard states and concentrations, all the $\mathfrak{E}$ and $d\mathfrak{E}/dT$ terms become $\mathfrak{E}^0$ and $d\mathfrak{E}^0/dT$ terms respectively. And if we make our measurements at 298°K, the above equations will then permit us to establish not only ΔG^0_{298} but also ΔH^0_{298} and ΔS^0_{298}.[44]

This potentiometric determination of ΔS^0_{298} offers us a golden opportunity to cross-check the corresponding value of ΔS^0_{298} obtained—with the aid of a Nernst heat theorem about which one may still have some doubt—from such purely thermal data as heat capacities, heats of fusion and volatilization, etc. The cross-check founded on these two completely independent determinations is an exceedingly searching test. For in the potentiometric method $d\mathfrak{E}^0/dT$, obtained as a small difference between much larger figures, is not nearly so well established as $\mathfrak{E}^0$ itself. And in the thermal method ΔS^0_{298} contains the entire accumulated error in measurements extending all the way back toward 0°K—which error is again much amplified because, like $d\mathfrak{E}^0/dT$, ΔS^0_{298} is established only as a comparatively small difference between much larger figures. All these difficulties notwithstanding, the agreement of the two determinations of ΔS^0_{298} is generally good and, for some reactions, almost unbelievably good.

▶ *Example 8*

In the galvanic cell symbolically represented as

$$\text{Ag (s), AgCl (s)}|\text{HCl (aq, 1 m)}|\text{Cl}_2\text{ (g, 1 atm), (Pt–Ir)},$$

the spontaneous net reaction is

$$\text{Ag (s)} + \tfrac{1}{2}\,\text{Cl}_2\text{ (g)} = \text{AgCl (s)}.$$

For this cell, at 298°K, Gerke gives $\mathfrak{E}^0 = 1.1362$ volts and $[d\mathfrak{E}^0/dT]_P = -5.95 \times 10^{-4}$ volt/°K. (a) What is the equilibrium constant for the cell reaction at 298°K? (b) How well does the potentiometric value of ΔS^0_{298} for the cell reaction agree with the thermal value $\Delta S^0_{298} = -13.85 \pm 0.25$ cal/°K?

Solution. (a) We can calculate the equilibrium constant directly from the quoted value for $\mathfrak{E}^0$.

$$\log K = \frac{n\mathfrak{F}\mathfrak{E}^0}{2.3RT} = \frac{n\mathfrak{E}^0}{0.0592} = \frac{(1)(1.1362)}{0.0592} = 19.21,$$

$$K = 1.6 \times 10^{19}.$$

(b) Direct substitution is again possible, yielding

$$\Delta S^0_{298} = n\mathfrak{F}\left[\frac{d\mathfrak{E}^0}{dT}\right]_P = (1)(96487)(-0.000595)$$

$$= -57.4 \text{ joules/°K}.$$

Converting to calories, and duly recognizing the limited accuracy of the temperature coefficient, we can express our result as

$$\Delta S^0_{298} = -13.73 \pm 0.10 \text{ cal/°K}.$$

Comparing this with the purely thermal value of -13.85 ± 0.25 cal/°K, we find ample basis for confidence in our experimental methods and in our use of the Nernst heat theorem. ◀

Temperature dependence of the equilibrium constant

Invariant as it is at constant temperature, the equilibrium constant K is well known to vary with temperature. The origin of this variability is easily identified. That ΔG^0 is itself a function of temperature is evident from equation (37), $\Delta G^0 = \Delta H^0 - T\,\Delta S^0$. Substituting for ΔG^0 from equation (59), we obtain

$$-RT \ln K = \Delta H^0 - T\,\Delta S^0,$$

$$\ln K = -\frac{\Delta H^0}{RT} + \frac{\Delta S^0}{R}. \tag{62}$$

Let us compare with each other the values of K at two temperatures spanned by an interval over which ΔH^0 may be regarded as constant. From equation (11) we learn that effective constancy of ΔH^0 requires that ΔC_P be negligibly small—under which condition the corresponding equation on p. 89 assures us that ΔS^0 must also be effectively constant. The last equation will then permit us to write for the two different temperatures:

$$\ln K_2 = -\frac{\Delta H^0}{RT_2} + \frac{\Delta S^0}{R},$$

$$\ln K_1 = -\frac{\Delta H^0}{RT_1} + \frac{\Delta S^0}{R}.$$

Subtracting the second equation from the first, we find

$$\ln K_2 - \ln K_1 = -\frac{\Delta H^0}{RT_2} + \frac{\Delta H^0}{RT_1},$$

$$\ln \frac{K_2}{K_1} = 2.303 \log \frac{K_2}{K_1} = -\frac{\Delta H^0}{R}\left(\frac{1}{T_2} - \frac{1}{T_1}\right). \tag{63}$$

This integral form of van't Hoff's law applies excellently to a great many systems, of which one is represented in Fig. 38.

▶ *Example 9*

At 25°C, 1.0×10^{-14} is the value of the ion-product constant for the dissociation $H_2O = H^+ + OH^-$. Given -13.8 kcal/mole as the heat of neutralization of 1 M NaOH with 1 M HCl, estimate the ion-product constant (K_0) for water at 0°C.

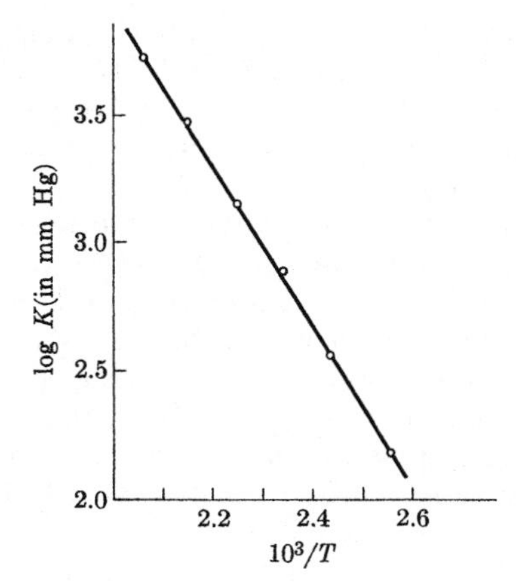

FIG. 38. Variation with temperature of the equilibrium constant for the vapor phase dissociation of acetic acid dimer, $(CH_3COOH)_2 = 2CH_3COOH$. [Data of Johnson and Nash, *J. Am. Chem. Soc.* **72**, 547 (1950).]

Solution. If we may take +13.8 kcal/mole as ΔH^0 for the dissociation reaction, and if we assume effective constancy of this value over the 25° span of temperature, substitution in the last equation will yield

$$\log \frac{1.0 \times 10^{-14}}{K_0} = -\frac{13800}{(2.303)(1.99)}\left(\frac{1}{298} - \frac{1}{273}\right)$$

$$= -3010(-0.000307) = +0.924,$$

$$\frac{1.0 \times 10^{-14}}{K_0} = 8.40,$$

$$K_0 = \frac{1.0 \times 10^{-14}}{8.40} = 1.2 \times 10^{-13}.$$

The empirical value of K_0 is 1.14×10^{-13}, so our assumptions are not too wide of the mark. ◀

From equation (62) we see that $-\Delta H^0/2.303R$ represents the slope of a plot of $\log K$ vs $1/T$. *If ΔH^0 is effectively constant* over the span of temperature concerned, the slope of the plotted line will be constant, and the line itself must then be straight. By actually measuring the line's slope, we can thus establish $-\Delta H^0/2.303R$, and thence ΔH^0. And equation (62) further indicates that, by determining the line's intercept with the vertical $\log K$ axis, we can also establish $\Delta S^0/2.303R$, and thence ΔS^0 for the reaction at issue. From values of K measured at several temperatures, this graphical analysis thus yields a value of ΔH^0 that can be compared with (or stand in lieu of) a direct calorimeteric determination, as well as a value of ΔS^0 that can be compared with (or stand in lieu of) a figure obtained by applying the Nernst heat theorem to measurements of heat capacities. Once again the cross-checks yield generally excellent agreement.[45]

Any significant variation of ΔH^0, over the temperature range concerned, will manifest itself in a curvature of the plot of log K *vs* $1/T$. At a given point on the plotted line, we can still construct a tangent; and from the tangent's slope and intercept we can, as before, determine ΔH^0 and ΔS^0 at the temperature to which the given point corresponds.[46] But the variation of ΔH^0 with temperature certainly *does* vitiate the assumption of constancy used in deriving equation (63). How shall we proceed when we can no longer use that integral equation to calculate the change of equilibrium constant with temperature? Problem 68 suggests several illuminating ways of deriving the requisite differential expression which, however, is most easily derived by calling again on the Gibbs-Helmholtz equation. For a change between standard states, this will be written

$$\frac{d}{dT}\left[\frac{\Delta G^0}{T}\right]_P = -\frac{\Delta H^0}{T^2}.$$

Substituting for ΔG^0 from equation (59), we find

$$\frac{d}{dT}\left[\frac{-RT\ln K}{T}\right]_P = -\frac{\Delta H^0}{T^2},$$

$$\frac{d\ln K}{dT} = +\frac{\Delta H^0}{RT^2}, \qquad (64)$$

where the constant-pressure restriction has been dropped because, as noted earlier, K is not a function of pressure.

Some qualitative indications suggested by Le Chatelier's principle are fully underwritten by equation (64). For an exothermic reaction, with ΔH^0 negative, the right side of that equation becomes negative, and the value of the equilibrium constant must then *decrease* with rise of temperature. For an endothermic reaction, with ΔH^0 positive, the right side of the equation will be positive, and the equilibrium constant must now *increase* with rise of temperature. But where such purely qualitative predictions represent the *most* we can draw from Le Chatelier's principle,[34] equation (64) invests us with power to make a fully quantitative evaluation of the change of equilibrium constant with temperature. When ΔH^0 is constant, integration of equation (64) duly yields equation (63), which may be solved as before. And when ΔH^0 is variable, we need only use equation (12) to express ΔH^0, in terms of ΔC_P, as a function of temperature. Substituting this function in equation (64), we obtain by integration an expression now rather more complicated than equation (63), but still an expression that renders calculable the change of equilibrium constant produced by a given change of temperature.

We saw earlier how values for ΔH^0 and K could be obtained from measurements of heat capacities and heats of reactions. We see now that, from values of ΔH^0 and K known for some one temperature, we can go on to calculate K at any other temperature to which our heat-capacity data extend. GIVEN PURELY THERMAL DATA, WE CAN CALCULATE THE EQUILIBRIUM CONSTANT OF ANY REACTION AT ANY TEMPERATURE—EVEN WHEN THE REACTION HAS NEVER BEEN ACHIEVED UNDER ANY CIRCUMSTANCES. This is the kind of calculation that guides and sustains what might otherwise seem forlorn endeavors to encompass what men have previously striven in vain to do. Thus, for example, thermodynamic considerations brought Nernst and Haber to an ammonia synthesis that had previously eluded the best efforts of such highly competent investigators as Ramsay and Le Chatelier. Today, more than half a century later, this first major technologic exploitation of thermodynamics probably remains that which most profoundly affects mankind. The first part of the relevant calculation has been given as Example 7; the last part will serve quite appropriately as our terminal illustrative example.

▶ *Example 10*

For the reaction

$$\tfrac{1}{2}\, N_2\ (g) + \tfrac{3}{2}\, H_2\ (g) = NH_3\ (g)$$

we found in Example 7 that, at 298°K, $\Delta H^0 = -11.02$ kcal and $K_p = 7.9 \times 10^2$. The formation of ammonia is thus strongly favored at room temperature, but at room temperature the rate of reaction is undetectably minute. Let it now be given that for N_2, H_2, and NH_3, at 298°K, the values of C_P are respectively 6.96, 6.89, and 8.38 cal/mole-°K. (a) What is the value of K_p for the ammonia-synthesis reaction at 500°C? (b) Under a total pressure of 100 atm at 500°C, what is the partial pressure of NH_3 at equilibrium with a stoichiometric mixture of 1 N_2 to 3 H_2?

Solution. (a) Whether we use equation (63) or equation (64) to determine K_p at 500°C (773°K) depends on whether or not ΔH^0 can be regarded as effectively constant over the temperature span 298–773°K. For the reaction that concerns us,

$$\Delta C_P = 8.38 - \tfrac{1}{2}(6.96) - \tfrac{3}{2}(6.89) = -5.43.$$

Even without the full calculation (given as Example 6 on p. 47) we see that, over the span of close to 500°K, this value of ΔC_P implies a change of more than 2 kcal in the magnitude of ΔH^0. Certainly we may *not* approximate as constant a value of ΔH^0 that actually changes by more than 20%. We must then turn to equation (64), and we begin by expressing ΔH^0, in terms of ΔC_P, as an explicit function of temperature. But now arises this awkward question: Will ΔC_P itself be constant over the span of 475°K? The answer is: Almost certainly not! Yet consider that ΔC_P is, after all, only a correction term. Fully to correct the correction may well be unnecessary when the change of temperature is only a few hundred degrees—and when, in any case, we propose to treat as ideal gases all too real at the relatively high pressure to which we must extend our calculation. Let us then treat ΔC_P as effectively constant: without at all changing the intrinsic character of the calculation, this approximation will much reduce its tediousness.

Using equation (12), we express ΔH^0 as a function of temperature:

$$\Delta H^0 = \Delta H^0_{298} + \Delta C_P(T - 298).$$

Substitution in equation (64) now yields

$$d \ln K = \frac{\Delta H^0_{298} + \Delta C_P(T - 298)}{RT^2}\, dT$$
$$= \frac{\Delta H^0_{298} - 298\, \Delta C_P}{RT^2}\, dT + \frac{\Delta C_P}{RT}\, dT.$$

When ΔC_P is treated as constant, integration over the range 298 to T°K involves only familiar simple integrals:

$$\int_{K_{298}}^{K_T} d \ln K = \frac{\Delta H^0_{298} - 298\, \Delta C_P}{R} \int_{298}^{T} \frac{dT}{T^2} + \frac{\Delta C_P}{R} \int_{298}^{T} \frac{dT}{T},$$
$$\ln \frac{K_T}{K_{298}} = -\frac{\Delta H^0_{298} - 298\, \Delta C_P}{R}\left(\frac{1}{T} - \frac{1}{298}\right) + \frac{\Delta C_P}{R} \ln \frac{T}{298}.$$

Shifting over to denary logarithms, and substituting numerical figures, we have

$$2.30 \log \frac{K_T}{K_{298}} = -\frac{-11{,}020 - 298(-5.43)}{1.99}\left(\frac{1}{T} - \frac{1}{298}\right) + \frac{-5.43}{1.99}\,(2.30) \log \frac{T}{298}.$$

We now divide through by 2.30, separate the log terms, and clean up a bit.

$$\log K_T - \log K_{298} = \frac{11{,}020 - 1620}{(1.99)(2.30)}\left(\frac{1}{T} - \frac{1}{298}\right) - 2.73 \log T + 2.73 \log 298.$$

Substituting the known values of $\log K_{298}$ and log 298, and doing a bit more cleaning up, we obtain

$$\log K_T - 2.90 = \frac{9400}{4.58}\left(\frac{1}{T} - \frac{1}{298}\right) - 2.73 \log T + (2.73)(2.48).$$

The rest is arithmetic, and the final relation is

$$\log K_T = \frac{2050}{T} - 2.73 \log T + 2.79.$$

We are now in a position to determine the equilibrium constant at 500°C, 773°K:

$$\log \mathrm{K}_{773} = \tfrac{2050}{773} - 2.73 \log 773 + 2.79 = -2.44 = -3 + 0.56,$$
$$\mathrm{K}_{773} = 3.6 \times 10^{-3}.$$

The value determined experimentally at 773°K is 3.8×10^{-3}. Approximating ΔC_P as constant, our extrapolation over the 475°-rise of temperature finds the equilibrium constant thus reduced to four-millionths of its former value. And our error in this evaluation of K_{773} is only 6%—surely not a bad showing for an approximate calculation.

(b) The value of K_{773} looks most unpropitious but, since two volumes of N_2 and H_2 combine to form one volume of NH_3, perhaps a reasonable yield of ammonia can be obtained by a synthesis under high pressure. We are asked to determine the equilibrium pressure of ammonia produced when a stoichiometric mixture of 1 N_2 to 3 H_2 is held at a constant total pressure of 100 atm at 773°K. If we symbolize by y the partial pressure of NH_3, $(100 - y)$ will represent the pressure in the equilibrium mixture of the N_2 and H_2 *together*. And, because we began with a stoichiometric mixture, at equilibrium the individual partial pressures must be $\frac{1}{4}(100 - y)$ for N_2 and $\frac{3}{4}(100 - y)$ for H_2. Using our calculated value for K_{773}, we substitute in the equilibrium expression to find

$$\frac{P_{NH_3}}{(P_{N_2})^{1/2}(P_{H_2})^{3/2}} = \frac{y}{[\frac{1}{4}(100 - y)]^{1/2}[\frac{3}{4}(100 - y)]^{3/2}} = 3.6 \times 10^{-3},$$
$$\frac{y}{(100 - y)^2} = \frac{3\sqrt{3}}{16} \times 3.6 \times 10^{-3} = 1.17 \times 10^{-3}.$$

This reduces to the following quadratic equation:

$$y^2 - 1.05 \times 10^3 y + 10^4 = 0.$$

Solving, we find the equilibrium partial pressure to be ~9.6 atm, compared with the experimental value of 10.4 atm (or 10.4%). Our doubly approximate method (taking ΔC_P as constant and treating real gases as ideal) thus yields a result amply good enough to support appraisal of the technologic feasibility of this NH_3 synthesis. ◀

Bibliographical Notes

1. On a putative "state principle," see beginning of paper by S. J. Kline and F. O. Koenig, *J. Appl. Mech.* **24,** 29 (1957).
2. H. T. Hall, *J. Chem. Educ.* **38,** 484 (1961).
3. For background and content of this selection of temperature scale and energy units, see H. F. Stimson, *Am. J. Phys.* **23,** 614 (1955).
4. M. W. Zemansky, *Temperatures Very Low and Very High* (Van Nostrand, 1964).
5. On the tremendous variety of calorimeters, see J. M. Sturtevant, "Calorimetry," Chap. 10 in *Physical Methods of Organic Chemistry*, ed. A. Weissberger (3rd ed., Vol. 1, Interscience, 1959); the historically oriented article by G. T. Armstrong, *J. Chem. Educ.* **41,** 297 (1964); and the up-to-date survey of instrumentation given by R. C. Wilhoit, *ibid.* **44,** A571, A629, A685, A853 (1967). On the "drop method" by which the ice calorimeter can be used to measure heat transfers at temperatures remote from 0°C, see for example D. C. Ginnings *et al.*, *Bur. Standards J. of Research* **45,** 23 (1950).
6. D. C. Ginnings and R. J. Corruccini, *ibid.* **38,** 583 (1947).
7. For a student determination of the latent heat of fusion of ice, see J. A. Soules, *Am. J. Phys.* **35,** 23 (1967).
8. W. H. Eberhardt, *J. Chem. Educ.* **41,** 483 (1964).
9. D. Kivelson and I. Oppenheim, "Work in Irreversible Expansions," *ibid.* **43,** 233 (1966).
10. S. D. Christian, "Reversible Work," *ibid.* **42,** 547 (1965).
11. For a compilation of the many sources that touch on "The Evolution of Energy Concepts from Galileo to Helmholtz," see Resource Letter EEC–1 by T. M. Brown, *Am. J. Phys.* **33,** 759 (1965); for interesting philosophical analyses, see D. W. Theobald, *The Concept of Energy* (Spon, 1966) and, on the scope and limitations of thermodynamics, K. G. Denbigh, *Chemistry in Britain* **4,** 338 (1968).
12. For a survey of thermochemical experiments potentially useful in the first-year chemistry course, see H. A. Neidig *et al.*, *J. Chem. Educ.* **42,** 26 (1965); for brief vignettes of the problems, procedures, and possi-

bilities offered by modern thermochemistry, see C. T. Mortimer and H. D. Springall, *Endeavour* **23,** 22 (1964), and R. A. Gilbert, *Record of Chemical Progress* **27,** 165 (1966).

13. For a historical account of G. H. Hess and his law, see H. M. Leicester, *J. Chem. Educ.* **28,** 581 (1951); on the need for caution in applying Hess's law, see T. W. Davis, *ibid.* **28,** 584 (1951), and G. P. Haight, *ibid.* **45,** 420 (1968).

14. The Born-Haber cycle represents one of the more obvious mergers of thermodynamics with the atomic-molecular viewpoint. On this cycle and its utility, see T. R. P. Gibb and A. Winnermann, *ibid.* **36,** 46 (1959); on the Kapustinskii equation and its affiliation with the Born-Haber cycle, see G. J. Moody and J. D. R. Thomas, *ibid.* **42,** 204 (1965), or the fully developed argument given by D. A. Johnson, *Some Thermodynamic Aspects of Inorganic Chemistry* (Cambridge Univ. Press, 1968).

15. The authoritative work on bond energies has long been T. L. Cottrell, *The Strengths of Chemical Bonds* (2nd ed., Butterworths, 1958); for a more recent survey, see the resource paper by S. W. Benson, *J. Chem. Educ.* **42,** 502 (1965); on the use of bond energies in the interpretation of descriptive chemistry, see R. A. Howald, *ibid.* **45,** 163 (1968).

16. That the ideal gas is insufficiently defined by its equation of state is an argument made convincingly by D. G. Miller and W. Dennis, *Am. J. Phys.* **28,** 796 (1960).

17. An excellent resource paper on "Combustion and Flame" is given by R. C. Anderson, *J. Chem. Educ.* **44,** 248 (1967); and on "Thermodynamics and Rocket Propulsion" see F. H. Verhoek, *ibid.* **46,** 140 (1969).

18. On the historical background of the second principle, see S. Carnot, *Reflections on the Motive Power of Fire,* ed. E. Mendoza (Dover, 1960), which also encompasses two major papers by Clapeyron and Clausius; and Chapter 7 "On the Physical View of Nature" in Vol. 2 of J. T. Merz, *A History of European Thought in the Nineteenth Century* (1912) (Dover, paperback ed., 1965).

19. Our representation of isothermal and adiabatic steps, in the Carnot cycle for an ideal gas, follows the usual policy of sacrificing accuracy of shape to gain clarity of representation. On the correct shape of the cycle, see P. Kirkpatrick, *Am. J. Phys.* **25,** 382 (1957).

20. On perpetual motion machines, see S. W. Angrist, *Sci. American* **218,** 114 (Jan. 1968); and, under the title "Something for Nothing," the article by D. E. H. Jones, *New Scientist* (Dec. 1965) 806.

21. On "Rudolf Diesel and His Rational Engine," see the article by L. Bryant, *Sci. American* **221,** 108 (August 1969); and, on fuel cells, the substantial article by J. Weissbart, *J. Chem. Educ.* **38,** 267 (1961).

22. On heat pumps see the presentation by J. F. Sandfort, *Sci. American* **184,** 54 (May 1951); on "The Stirling Refrigeration Cycle" see the excellent article so entitled by J. W. L. Kohler, *ibid.* **212,** 119 (April 1965).
23. Amply fascinating, though little relevant to chemistry, the existence in certain very special circumstances of systems assigned negative temperatures is discussed by C. E. Hecht, *J. Chem. Educ.* **44,** 124 (1967).
24. On the thermodynamics of life and its processes, see the outdated but still stimulating tract by E. Schrödinger, *What is Life?* (Cambridge Univ. Press, 1945). For an account both more up to date and more technical, see A. L. Lehninger, *Bioenergetics* (Benjamin, 1965); or the somewhat sophisticated monograph by H. J. Morowitz, *Energy Flow in Biology* (Academic Press, 1968). On the more limited topic of energy conservation in biological oxidations, see J. Kirschbaum, *J. Chem. Educ.* **45,** 28 (1968).
25. On "spread" as a verbalism superior to others, see E. A. Guggenheim, *Research* **2,** 450 (1949). On how verbalisms can be replaced by formalisms, see J. S. Dugdale, *Entropy and Low Temperature Physics* (Hutchinson, London, 1966); or L. K. Nash, *Elements of Statistical Thermodynamics* (Addison-Wesley, 1968).
26. On entropies of melting, see A. R. Ubbelohde, *Quart. Rev.* **4,** 356 (1950).
27. On the statistical mechanical basis of Trouton's rule, see D. McLachlan, Jr. and R. J. Marcus, *J. Chem. Educ.* **34,** 460 (1957).
28. On the derivation of the Nernst heat theorem see E. M. Loebl, *ibid.* **37,** 361 (1960); and D. Chandler and I. Oppenheim, *ibid.* **43,** 525 (1966). On the virtues and defects of the "third law of thermodynamics" see F. E. Simon, *Physica* **4,** 1089 (1937); and also J. Wilks, *The Third Law of Thermodynamics* (Oxford Univ. Press, 1961).
29. On the great simplicity of the analytical extrapolation using the Debye T^3-law, see G. Calingaert, *J. Chem. Educ.* **31,** 487 (1954).
30. On the great importance of low-temperature heat-capacity measurements, see E. F. Westrum, Jr., *ibid.* **39,** 443 (1962).
31. The entropies of dissolved substances, and the entropy changes in composite reactions in solution, are discussed by E. L. King, *ibid.* **30,** 71 (1953), **43,** 478 (1966). The conventions used in defining the entropies of dissolved species are thoroughly analyzed by R. M. Noyes, *ibid.* **40,** 2, 116 (1963); the far-reaching effects of different choices of standard state are illustrated for the so-called chelate effect by H. A. Bent, *J. Phys. Chem.* **60,** 123 (1956) and by J. E. Prue, *J. Chem. Educ.* **46,** 12 (1969).
32. On how the free-energy criterion is used to determine the feasibility of a chemical reaction, see G. E. MacWood and F. H. Verhoek, *ibid.*

38, 334 (1961); for the closely related analysis of the stability of compounds see J. A. Allen, *ibid.* **39,** 561 (1962).

33. The occurrence of spontaneous endothermic reactions, as the result of entropy increases consequent to a large net increase in the moles of gas present, is well illustrated by G. W. J. Matthews, *ibid.* **43,** 476 (1966).
34. The relation of the triple point to the normal freezing point of air-saturated water is clarified by F. L. Swinton, *ibid.* **44,** 541 (1967).
35. On Le Chatelier's principle, see J. de Heer, *ibid.* **34,** 375 (1957); and L. Katz, *ibid.* **38,** 375 (1961).
36. An illuminating analysis of the linearity of ln P *vs* $1/T$ plots is given by O. L. I. Brown, *ibid.* **28,** 428 (1951), who suggests a good derivation of the Clausius-Clapeyron equation. On the several substantial corrections that must be made in calculating $\Delta\overline{H}_{vap}$ from the slope of ln P *vs* $1/T$ plots, see L. Brewer and A. W. Searcy, *ibid.* **26,** 548 (1949). An interesting calculation of the vapor pressure of water, based on the condition that equilibrium in an isolated system must represent an entropy maximum, is given by M. H. Everdell, *ibid.* **46,** 107 (1969).
37. On the virtues of "Fixed Pressure Standard States," see the paper so entitled by P. A. Rock, *ibid.* **44,** 104 (1967).
38. For a full analysis of the nature and magnitude of the difference, see A. G. Williamson, *ibid.* **43,** 211 (1966); and V. Fried, *ibid.* **45,** 720 (1968).
39. Another such case is well illustrated in a student experiment on the solubility of diphenylamine in naphthalene: see B. H. Mahan, *ibid.* **40,** 293 (1963).
40. On modern instrumentation for osmometry, see P. F. Lott and F. Millich, *ibid.* **43,** A191, A295 (1966); and see also the extended theoretical analysis given in J. F. Thain's *Principles of Osmotic Phenomena* (No. 13 in the series of Royal Inst. of Chem. Monographs for Teachers, 1967).
41. For the alternative definition of the ideal gas mixture, in terms of some of these consequences, see P. J. Robinson, *ibid.* **41,** 654 (1964); and also the somewhat related analysis by M. L. Kremer, *ibid.* **42,** 649 (1965).
42. The great simplicity of the thermodynamic calculation of solubility products is indicated by W. H. Waggoner, *ibid.* **35,** 339 (1958); the balance of enthalpy and entropy effects in determining the values of ionization constants for weak acids is examined by C. R. Allen and P. G. Wright, *ibid.* **41,** 251 (1964), and by M. J. Frazer, *Educ. in Chem.* **1,** 39 (1964).

43. The classic analysis is that of W. M. Latimer, *Oxidation Potentials*, 2nd ed., Prentice-Hall, 1952. For a new conception of electrode potentials, permitting easy recognition of the thermodynamic origin of galvanic voltages, see J. B. Ramsey, *J. Chem. Educ.* **38,** 353 (1961).
44. For two student experiments that permit intercomparison of calorimetric and potentiometric determinations of heats of reaction, see D. L. Hill *et al.*, *ibid.* **42,** 541, 544 (1965); for a student experiment, on cell potentials, offering an illustration of the usefulness of the Gibbs-Helmholtz relation, see R. S. Johnson and D. E. Crawford, *ibid.* **46,** 52 (1969). For discussion of the value of potentiometric measurements in thermodynamic studies of aqueous solutions at elevated temperatures, see M. H. Lietzke and R. W. Stoughton, *ibid.* **39,** 230 (1962).
45. M. J. Joncich *et al.*, *ibid.* **44,** 598 (1967) describe a straightforward isoteniscopic experiment that yields the student figures for ΔG^0, ΔH^0, and ΔS^0 in the decomposition of ammonium carbamate.
46. On this aspect of the van't Hoff equation, see J. T. MacQueen, *ibid.* **44,** 755 (1967).

Appendix II

*Some Operations of the Calculus**

To follow completely all the derivations in this book, you need grasp only a very few of the most elementary concepts and relations of the calculus. All these mathematical tools are developed in the present appendix, in which we represent a general "function" as $y = f(x)$, read: "y equals f of x." This does not signify that y is f times x, but simply that y is so expressed in terms of x that, when x is given, the value of y is also determined. Some simple functions are $y = 2x$, $y = 1/x$, $y = \sin x$.

Differentiation. Consider the changing velocity of an automobile accelerating, from a standing start, along a straight road. Figure A–1 shows the velocity plotted as a function of time.

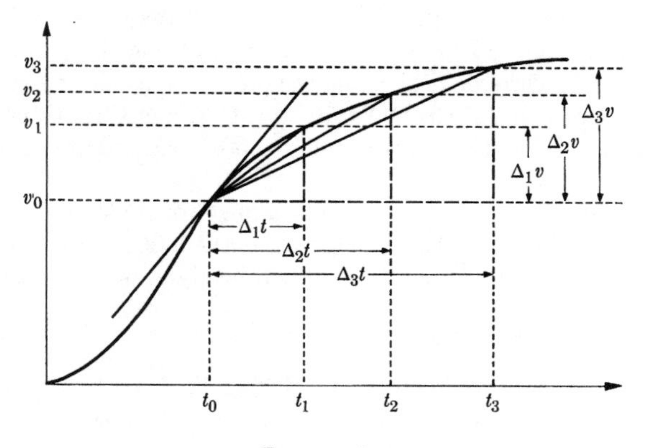

FIGURE A–1

Suppose we wish to know the acceleration—that is, the rate at which the velocity is changing—at time t_0. Experimentally we might read the

* For a full, and impeccable, account of the arguments sketched in this appendix see G. B. Thomas, *Calculus and Analytic Geometry*, 4th ed., *passim.*

velocity once at time t_0 and again at some later time t_3. Then

$$\frac{v_3 - v_0}{t_3 - t_0} = \frac{\Delta_3 v}{\Delta_3 t} = \alpha_3$$

and we may say that this quotient represents the rate of change of the velocity at time t_0. But clearly this is *not the instantaneous* acceleration at time t_0—represented by the slope (i.e., the vertical "rise" over the horizontal "run") of the tangent to the curve at point (t_0, v_0)—but rather the *average* acceleration over the period t_0 to t_3 represented by the slope of the secant line from (t_0, v_0) to (t_3, v_3). We come closer to the instantaneous acceleration represented by the tangent line if we make our terminal velocity measurement at time t_2, and still closer if we can make the measurement at time t_1, but clearly we will get an accurate value only when the time interval becomes infinitesimal, in which case the second velocity will differ infinitesimally (and wholly undetectably) from the first. How then are we to proceed?

We want the value of $\Delta v/\Delta t$ as $\Delta t \rightarrow 0$ (read "delta t approaches zero"). This value of the instantaneous acceleration we cannot get directly from readings of stopwatch and speedometer. However, given such readings, we can prepare a graph like that of Fig. A–1. By actually constructing the tangent line, we can determine its slope and take this as the instantaneous acceleration. There is nothing wrong in principle with this somewhat clumsy procedure, and on some occasions it may be the only option open to us. But when v can be expressed as an analytical function of t, the differential calculus offers a much more elegant approach to the problem.

Suppose that $v = mt - nt^2$, where m and n are constants. At the instant t_0, the velocity $v_0 = mt_0 - nt_0^2$. After the lapse of the further time interval Δt the velocity at time $t_0 + \Delta t$ will be given by the equation

$$\begin{aligned} v_0 + \Delta v &= m(t_0 + \Delta t) - n(t_0 + \Delta t)^2 \\ &= m(t_0 + \Delta t) - n(t_0^2 + 2t_0\,\Delta t + \overline{\Delta t}^2). \end{aligned}$$

To find the actual change of velocity, Δv, in the time interval Δt, we subtract from the last equation the expression for the initial condition:

$$v_0 = mt_0 - nt_0^2.$$

We have then

$$\Delta v = m\,\Delta t - n(2t_0\,\Delta t + \overline{\Delta t}^2).$$

Now we are interested in the limiting value of the ratio $\Delta v/\Delta t$, and to form the ratio we divide through the last equation by Δt, to get

$$\frac{\Delta v}{\Delta t} = m - n(2t_0 + \Delta t).$$

As such, the ratio $\Delta v/\Delta t$ gives us the slope not of the tangent line but of a secant; but as $\Delta t \to 0$ the value of this ratio approaches as a limit the slope of the tangent line. What is that slope? Looking at the last equation, we see that any term containing Δt in the numerator will go to zero as $\Delta t \to 0$. Hence

$$\lim_{\Delta t \to 0} \left[\frac{\Delta v}{\Delta t}\right] = \lim_{\Delta t \to 0} [m - n(2t_0 + \Delta t)] = m - 2nt_0.$$

This is the expression for the slope of the tangent line at t_0. Since t_0 can be *any* point on the curve, we have only to insert any particular value of t_0 to read off from the last formula the instantaneous acceleration at that time. And to the term

$$\lim_{\Delta t \to 0} \left[\frac{\Delta v}{\Delta t}\right]$$

we assign the special symbol dv/dt (read, "the derivative of v with respect to t"):

$$\frac{dv}{dt} = m - 2nt.$$

This somewhat tedious procedure for determining the slope can be generalized, so that one need not go through the entire procedure in other similar cases. Consider what we have done:

1. We wrote down the function $v = f(t)$.
2. We formulated the expression $v + \Delta v = f(t + \Delta t)$.
3. We subtracted the first expression from the second to get $\Delta v = f(t + \Delta t) - f(t)$.
4. We divided through by Δt to get

$$\frac{\Delta v}{\Delta t} = \frac{f(t + \Delta t) - f(t)}{\Delta t}.$$

   The division by Δt can only be indicated when we speak of the general function, $f(t)$, but would actually be carried through, as in the example, when dealing with any specific function.

5. We took the limit as $\Delta t \to 0$, and so arrived at the derivative:

$$\frac{dv}{dt} = \lim_{\Delta t \to 0} \left[\frac{f(t + \Delta t) - f(t)}{\Delta t}\right].$$

This is the *general* procedure for determining an instantaneous rate of change. [*Note:* For Newton, what we call the calculus was the method of fluxions, from L. *fluxio(-onis)*, a flowing.] There is, of course, no reason to restrict the variables to v and t, and for the general function $y = f(x)$ we write a similar expression for the instantaneous rate of change of y with x:

$$\frac{dy}{dx} = \lim_{\Delta x \to 0} \left[\frac{f(x + \Delta x) - f(x)}{\Delta x}\right]. \qquad \textit{Definition}$$

For the bracketed expression we will in the future use the more compact expression $f'(x)$ (read "f prime of x"), so that when $y = f(x)$ then $dy/dx = f'(x)$.

Given the above definition, we can at once write down the derivatives of a great many simple functions.

$$\text{If } y = x^3; \text{ then } \frac{dy}{dx} = \lim_{\Delta x \to 0} \left[\frac{(x^3 + 3x^2\,\Delta x + 3x\,\overline{\Delta x}^2 + \overline{\Delta x}^3) - x^3}{\Delta x}\right]$$
$$= \lim_{\Delta x \to 0} [3x^2 + 3x\,\Delta x + \overline{\Delta x}^2] = 3x^2.$$

$$\text{If } y = x^4; \text{ then } \frac{dy}{dx} = \lim_{\Delta x \to 0} \left[\frac{(x^4 + 4x^3\,\Delta x + 6x^2\,\overline{\Delta x}^2 + 4x\,\overline{\Delta x}^3 + \overline{\Delta x}^4) - x^4}{\Delta x}\right]$$
$$= \lim_{\Delta x \to 0} [4x^3 + 6x^2\,\Delta x + 4x\,\overline{\Delta x}^2 + \overline{\Delta x}^3] = 4x^3.$$

These results, together with that earlier obtained in the illustrative example, suffice to suggest what can indeed be shown to be a general formula. For all integral values of n, positive or negative:

$$\text{If } y = x^n, \qquad \text{then } \frac{dy}{dx} = nx^{n-1}.$$

The illustrative example suggests also two further conclusions the (easy) proofs of which are omitted. First,

$$\text{If } y = x^m + x^n, \qquad \text{then } \frac{dy}{dx} = mx^{m-1} + nx^{n-1}.$$

And, more generally still,

$$\text{If } y = f(x) + g(x), \qquad \text{then } \frac{dy}{dx} = f'(x) + g'(x).$$

Second, where C symbolizes any constant term,

$$\text{If } y = Cx^n, \qquad \text{then } \frac{dy}{dx} = (C)(n)x^{n-1}.$$

And, much more generally still,

$$\text{If } y = Cf(x), \qquad \text{then } \frac{dy}{dx} = Cf'(x).$$

These formulas give us an easy means of finding the slopes of curves—and, much more generally, the rate of change of one variable with another in a great many cases of practical interest. For our work in thermodynamics we need only three more formulas. The first is almost trivial. Suppose that $y = C$. This is the equation of a straight line parallel to the x-axis, and so with slope zero at all points. The use of the basic definition further confirms the conclusion that

$$\text{If } y = C, \qquad \text{then } \frac{dy}{dx} = 0.$$

The next relation is simply announced, not derived. We are most used to denary logarithms that satisfy the definition $x = 10^{\log x}$. Another set of logarithms is founded on the base $e = 2.718\ldots$, and satisfies the definition $x = e^{\ln x}$. The two sets of logarithms are related by the equation $2.303 \log x = \ln x$. Logarithms to the base e are far less convenient for, say, ordinary trigonometric calculations, but they have important mathematical properties, one of which is

$$\text{If } y = \ln x, \qquad \text{then } \frac{dy}{dx} = \frac{1}{x}\cdot$$

One last relation:

$$\text{If } y = f(x) \cdot g(x), \qquad \text{then } \frac{dy}{dx} = f(x) \cdot g'(x) + g(x) \cdot f'(x).$$

This we can prove with the aid of the basic definition. Letting $f(x) = u$ and $g(x) = v$, we rewrite the function in the form $y = uv$. Now, by definition,

$$\begin{aligned}\frac{dy}{dx} &= \lim_{\Delta x \to 0} \left[\frac{(u + \Delta u)(v + \Delta v) - uv}{\Delta x}\right] \\ &= \lim_{\Delta x \to 0} \left[\frac{uv + u\,\Delta v + v\,\Delta u + \Delta u\,\Delta v - uv}{\Delta x}\right] \\ &= \lim_{\Delta x \to 0} \left[u\frac{\Delta v}{\Delta x} + v\frac{\Delta u}{\Delta x} + \Delta u\frac{\Delta v}{\Delta x}\right]\cdot\end{aligned}$$

But when Δx approaches zero so will Δu and Δv. For consider that

$$\lim\,(\Delta u) = \lim\left(\frac{\Delta u}{\Delta x}\,\Delta x\right) = \lim\left(\frac{\Delta u}{\Delta x}\right)\lim\,(\Delta x) = \left(\frac{du}{dx}\right)(0) = 0.$$

Consequently it follows that

$$\frac{dy}{dx} = u\frac{dv}{dx} + v\frac{du}{dx} + 0\frac{dv}{dx} = u\frac{dv}{dx} + v\frac{du}{dx}\cdot$$

But by definition

$$u = f(x) \quad \text{and} \quad \frac{du}{dx} = f'(x),$$

and also

$$v = g(x) \quad \text{and} \quad \frac{dv}{dx} = g'(x).$$

Substituting in the previous equation we have then what we set out to prove:

$$\frac{dy}{dx} = f(x) \cdot g'(x) + g(x) \cdot f'(x).$$

The derivative dy/dx, as so far considered, is a single symbol and not a fraction. But when x is an independent variable and y is a function of x, we can attach meaning to dy and dx separately. For the function $y = f(x)$ we have

$$\frac{dy}{dx} = f'(x).$$

Then dx ("the differential of x") is to be regarded as an infinitesimal increment in x, and dy ("the differential of y") is a function of x and dx given by

$$dy = f'(x)\, dx.$$

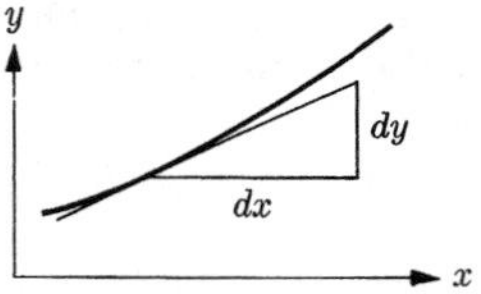

FIGURE A–2

That is, dy/dx is the *rate* at which y changes per unit change in x, and then dy is the *amount* by which y changes (measured along the tangent to the curve) per unit change in x. If dx is held to infinitesimal values, the tangent line diverges only infinitesimally from the curve, and dy becomes simply the infinitesimal increment in y corresponding to the infinitesimal increment dx.

Function	Derivative	Differential	Example
$y = C$	$\frac{dy}{dx} = \frac{d(C)}{dx} = 0$	$dy = d(C) = 0$	$d(C) = 0$
$y = x^n$	$\frac{dy}{dx} = \frac{d(x^n)}{dx} = nx^{n-1}$	$dy = d(x^n) = nx^{n-1}\,dx$	$d(x^{-1}) = -\frac{dx}{x^2}$
$y = \ln x$	$\frac{dy}{dx} = \frac{d(\ln x)}{dx} = \frac{1}{x}$	$dy = d(\ln x) = \frac{dx}{x}$	$d(\ln x) = \frac{dx}{x}$
$y = Cu$	$\frac{dy}{dx} = \frac{d(Cu)}{dx} = C\frac{du}{dx}$	$dy = d(Cu) = C\,du$	$d(12x^2) = 24x\,dx$
$y = u + v$	$\frac{dy}{dx} = \frac{d(u+v)}{dx} = \frac{du}{dx} + \frac{dv}{dx}$	$dy = d(u+v) = du + dv$	$d(x^2 + x^3) = 2x\,dx + 3x^2\,dx$
$y = uv$	$\frac{dy}{dx} = \frac{d(uv)}{dx} = u\frac{dv}{dx} + v\frac{du}{dx}$	$dy = d(uv) = v\,du + u\,dv$	$d(x \ln x) = \ln x\,dx + dx$

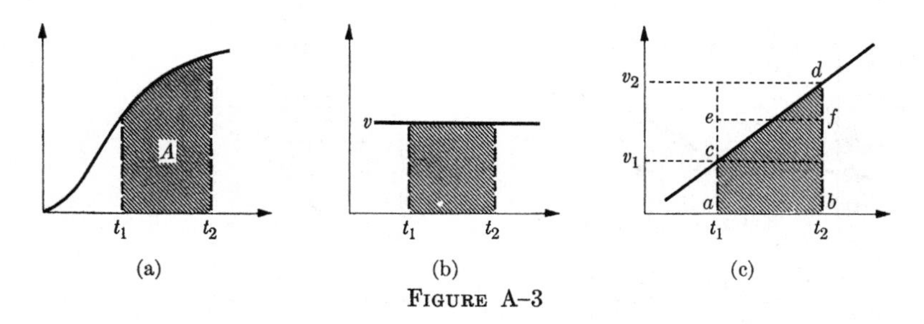

FIGURE A–3

Integration. The plot of velocity, as a function of time, for a car accelerating from rest is again depicted in Fig. A–3(a). We now raise this question: what is the total *distance* covered by the car in any period of time, say in the interval t_1 to t_2? Now there are two cases in which this question poses no difficulty. If in the interval t_1 to t_2 the car proceeds at constant velocity, as shown in Fig. A–3(b), then the distance traversed is simply the product of the velocity and the elapsed time, or $v(t_2 - t_1)$. Note that this product is just the shaded area in Fig. A–3(b). An only slightly less simple case is that in which velocity changes uniformly with time, so that the plot of v against t is a line of constant slope, as shown in Fig. A–3(c). Here we easily find the distance traversed by forming the product of the *average* velocity and the time interval, or

$$\frac{v_1 + v_2}{2}(t_2 - t_1).$$

By so doing, we determine the area $\overline{aefb}$ which is equal to the shaded area $\overline{acdb}$ enclosed between the velocity plot, the x-axis, and the verticals corresponding to the limits t_1 and t_2.

Returning now to the original problem sketched in Fig. A–3(a), we see that here too the shaded area will correspond to the distance sought. Consider that we divide the period t_1 to t_2 into a number of equal short increments of time, Δt. Although the car's velocity changes continuously as a function of time, for the short interval $\Delta_i t$ we may think to approximate the velocity, as shown in Fig. A–4(a), as the initial (minimum) velocity ${}_m v_i$ for that interval. The distance covered in the time interval is then ${}_m v_i\,(\Delta_i t)$, which is simply the area of one rectangular block. For the total distance covered we will, according to this mode of reckoning, add the areas of all the rectangular blocks falling between t_1 and t_2. A compact symbolization of this operation is $\sum_{i=1}^{5} {}_m v_i\,(\Delta_i t)$, where the Greek sigma instructs us to make the indicated summation of all the individual terms having the indicated form. And it seems intuitively evident that as the standard interval $\Delta t \to 0$ and as the number n of ever-narrower blocks is increased without limit, this sum will approach as a limit the area enclosed

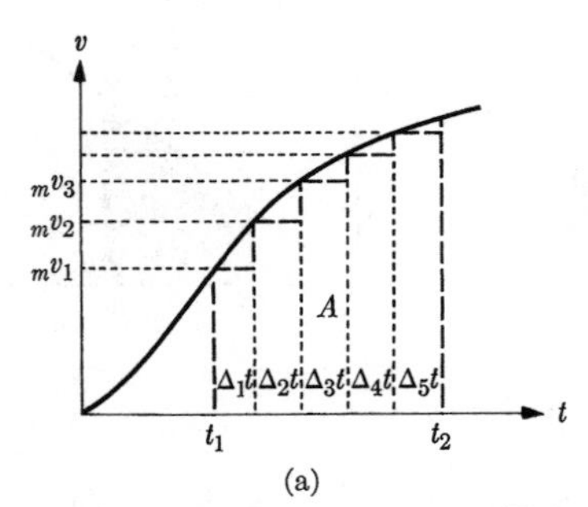

(a)

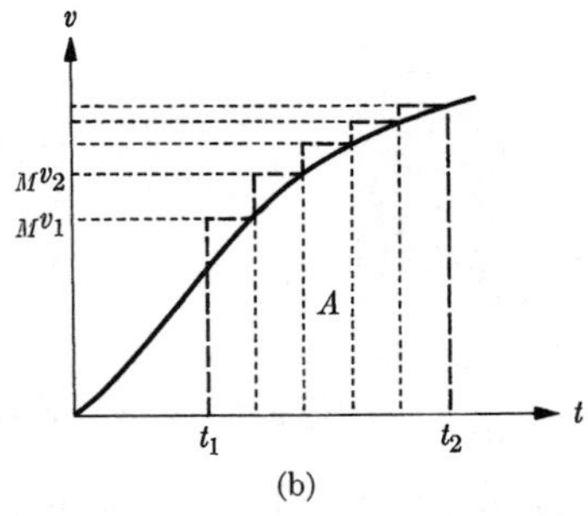

(b)

FIGURE A–4

under the curve between the verticals at t_1 and t_2. That is,

$$A = \lim_{\Delta t \to 0} \sum_{i=1}^{n} {}_m v_i \, (\Delta_i t).$$

One may perhaps have an uneasy feeling that this mode of approach yields only an *approximation*, a minimum value for the area under the curve. But consider that we may also set out, as shown in Fig. A–4(b), from the final (maximum) velocity, ${}_M v_i$, attained in each time interval. For the total distance traveled we have then the approximation

$$\sum_{i=1}^{5} {}_M v_i \, (\Delta_i t).$$

And if now again we shorten the time interval, and so increase without limit the number of blocks, n, it seems clear that in this case too we can write

$$A = \lim_{\Delta t \to 0} \sum_{i=1}^{n} {}_M v_i \, (\Delta_i t).$$

That is, whether we approach from the direction of an *over*estimate or from the direction of an *under*estimate, in the limit we approach the same area. Given a plot of velocity as a function of time we can then determine, right on the graph, the distance traversed in the period t_1 to t_2 simply by counting up squares in the enclosed region. This geometric approach—perfectly effective, and often unavoidable—falls in the same category of (crude) operations as the determination of a slope by actual construction of a tangent line. In the latter case we found a more elegant alternative, the analytical operation of differentiation, available whenever the plotted curve can be expressed as a function $y = f(x)$. In exactly the same way, *integration* furnishes us with an analytical method for the determination of areas enclosed under curves that can be expressed in the functional form $y = f(x)$.

Let us consider the general case of a changing velocity expressible as $v = f(t)$, making no assumption of a constantly increasing velocity. Taking some standard short time interval Δt, we ask what will be the distance ΔA covered in the period between some particular time t and the time $t + \Delta t$. We take as the velocity *throughout* this time interval the value $v = f(t)$, and it makes no difference whether this is a maximum, a minimum, or an intermediate value for the velocity in the indicated period. For, as we have just seen, as $\Delta t \to 0$ the maximum and minimum values of v in the ever-decreasing time interval are simply "squeezed" towards each other and towards the value $v = f(t)$. As an approximation for the distance covered we have then the thin rectangular slice cut out by the vertical boundaries at t and $t + \Delta t$, with area $\Delta A = v\,\Delta t = f(t) \cdot \Delta t$. And as a rigorous result we have, in the limit as $\Delta t \to 0$, that

$$\lim_{\Delta t \to 0} \left[\frac{\Delta A}{\Delta t}\right] = \frac{dA}{dt} = f(t).$$

Now we are seeking the *entire* distance covered in the period t_1 to t_2—the entire area enclosed under the plot between the vertical boundaries at t_1 and t_2. We seek then some relation of the form $A = F(t)$ from which we can evaluate, as a function of time, the total distance concerned. Given a relation we write in differential form as $dA = f(t)\,dt$, we seek the desired relation. Clearly, we must somehow "dis-differentiate" or "undifferentiate." This operation we call integration and symbolize $\int$. We write then

$$\int dA = \int f(t)\,dt.$$

Bidden by the integral sign, we ask ourselves what function of A will, when differentiated, yield the term dA. The answer is surely obvious, and we then write

$$A = \int f(t)\,dt.$$

What about the right side of the equation? We see that the $F(t)$ we seek must be that function of t which on differentiation yields the term $f(t)\,dt$. That is, $f(t) = F'(t)$. That integration is just the inverse of differentiation we can then symbolize as follows:

$$\int f(t)\,dt = \int F'(t)\,dt = \int dF(t) = F(t).$$

Given $f(t)$, we hope to establish $F(t)$. Take the particular case with which we began—that of an accelerating car for which $v = mt - nt^2$. We then write

$$A = \int f(t)\,dt = \int (mt - nt^2)\,dt.$$

What function of t will, when differentiated, yield $(mt - nt^2)\,dt$? Try the first term first:

Differentiation of t^2 yields $2t\,dt$.
Differentiation of $\frac{1}{2}t^2$ yields $t\,dt$.
Differentiation of $(m)(\frac{1}{2})t^2$ yields $mt\,dt$.

A similar development yields the further conclusion that

Differentiation of $(n)(\frac{1}{3})t^3$ yields $nt^2\,dt$.

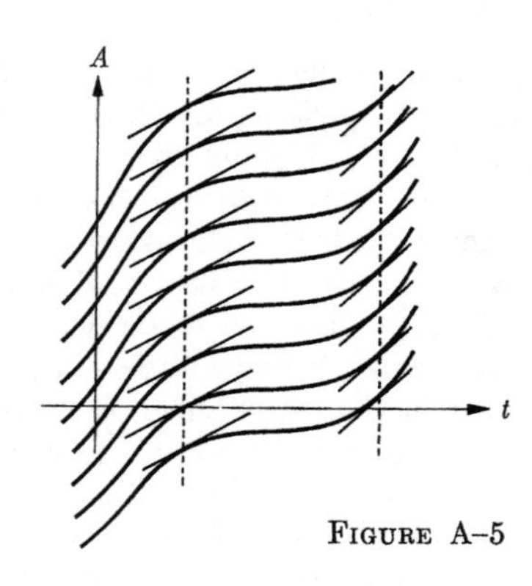

FIGURE A–5

Hence we can at last write

$$A = \frac{mt^2}{2} - \frac{nt^3}{3}, \tag{a}$$

and our problem is at last solved. Or is it?

Consider again the differential equation $dA/dt = f(t)$. Observe that it can correspond to any of the whole family of curves sketched in Fig. A–5. All the equation tells us is that the *slope* of a plot of A against t is a particular function of t, and any of an infinite number of curves will meet this requirement. By re-examining the last derivation, we see that in fact we have oversimplified: we do *not* have a unique solution for A. The equation

$$dA = (mt - nt^2)\,dt$$

can be obtained not only from equation (a) but also from all equations of the form

$$A = \frac{mt^2}{2} - \frac{nt^3}{3} + C,$$

where C can be *any constant* since, as earlier noted, $d(C) = 0$. For different values of C the last equation produces the set of parallel curves shown in Fig. A–5. For any given value of t, A may then assume any one of the infinite possibilities corresponding to any one of the infinite values that C may take. Our problem is then *not* fully solved; we must somehow establish along which one of the curves $A = F(t) + C$ we are to read.

We easily meet this apparently difficult problem by a specification of "limits"—corresponding to the vertical lines bounding the interval in question. Suppose, for example, we wish to determine the distance traveled by the car in the period beginning at time 0, when it starts from rest, and ending at some later time t. At the "lower" (beginning) limit, when $t = 0$, the distance traversed will be $A = 0$, and we seek the value of A at time t.

We display the limits in the form

$$\int_0^A dA = \int_0^t (mt - nt^2)\, dt. \tag{b}$$

This specification of limits at once establishes a choice of one of the curves $A = F(t) + C$: we must select the curve running through the origin ($A = 0$ when $t = 0$), and along this curve we can then read off the value of A corresponding to any specified time t.

The geometric operation here described in words is easily reduced to a rigorously specifiable analytical operation. Treating the general case, suppose we wish to know the distance traveled by the car in the time between t_1 and t_2. Let A_1 represent the distance (relative to some arbitrary standard) at time t_1; let A_2 represent the distance at time t_2. What we then want to find is $A_2 - A_1$. We begin by substituting the lower set cf limits to establish the value of the constant C:

$$A_1 = \frac{mt_1^2}{2} - \frac{nt_1^3}{3} + C,$$

whence

$$C = A_1 - \left(\frac{mt_1^2}{2} - \frac{nt_1^3}{3}\right).$$

Having obtained this value of C, we can now demand that at the "upper" limit

$$A_2 = \frac{mt_2^2}{2} - \frac{nt_2^3}{3} + \left\{A_1 - \left(\frac{mt_1^2}{2} - \frac{nt_1^3}{3}\right)\right\}.$$

This, on rearrangement, yields

$$A_2 - A_1 = \left(\frac{mt_2^2}{2} - \frac{nt_2^3}{3}\right) - \left(\frac{mt_1^2}{2} - \frac{nt_1^3}{3}\right).$$

Here is an expression for the distance traveled in the period t_1 to t_2. At last we have solved our problem.

The somewhat clumsy procedure in which we solve, and substitute, for C is easily avoided. Returning to equation (b), we integrate and indicate the limits thus:

$$A\Big]_{A_1}^{A_2} = \frac{mt^2}{2} - \frac{nt^3}{3}\Big]_{t_1}^{t_2}.$$

On each side of the equation we simply subtract from the value of the function at the upper limit, the value of the function at the lower limit, and so at once obtain the relation previously secured more laboriously:

$$A_2 - A_1 = \left(\frac{mt_2^2}{2} - \frac{nt_2^3}{3}\right) - \left(\frac{mt_1^2}{2} - \frac{nt_1^3}{3}\right).$$

And now we may see our way to a general formulation of the operation of integration. Any time we can relate the change of one variable to that of another through an equation of the type $dA = f(x)\,dx$, then we can always conclude that, for the limits (x_1, A_1) and (x_2, A_2), it is the case that

$$A_2 - A_1 = F(x_2) - F(x_1),$$

where $F(x)$ is the function that, on differentiation, yields the function $f(x)\,dx$.

We close with a few specific formulas for the determination of $F(x)$ from $f(x)$. We have already encountered examples of the general relation

$$\int_{x_1}^{x_2} x^m\,dx = \frac{x^{m+1}}{m+1}\Bigg]_{x_1}^{x_2}$$

and also of

$$\int_{x_1}^{x_2} [f(x)\,dx + g(x)\,dx] = F(x) + G(x)\Big]_{x_1}^{x_2}.$$

Two more formulas follow, just as these first two do, directly from the tabulation on page 150. The first is

$$\int_{x_1}^{x_2} \frac{dx}{x} = \ln x\Big]_{x_1}^{x_2} = \ln x_2 - \ln x_1 = \ln\frac{x_2}{x_1}.$$

The second, often a very helpful relation, indicates that integrals can always be simplified by moving constants, but *only* constants, across the integral sign:

$$\int_{x_1}^{x_2} Cf(x)\,dx = C\int_{x_1}^{x_2} f(x)\,dx = CF(x)\Big]_{x_1}^{x_2} = C[F(x_2) - F(x_1)].$$

One last relation is of a different sort, though scarcely less obvious than the foregoing. What happens when we "invert limits"? We then have

$$\int_{x_1}^{x_2} f(x)\,dx = F(x_2) - F(x_1), \qquad \int_{x_2}^{x_1} f(x)\,dx = F(x_1) - F(x_2),$$

whence it follows that

$$\int_{x_1}^{x_2} f(x)\,dx = -\int_{x_2}^{x_1} f(x)\,dx.$$

Appendix III

Problems

Standard gravitational acceleration, $g = 9.807$ meter/sec^2

Standard atmosphere = 76 cm Hg = 1.033 kgm/cm^2 = 10.13 newton/cm^2

Ideal gas law constant, R = 1.987 cal/mole-°K
= 8.314 joule/mole-°K
= 0.08205 lit-atm/mole-°K

Electrochemical faraday, $\mathfrak{F}$ = 96487 coulombs/gm-equivalent

0°C = 273.15°K $\ln X = 2.303 \log X$

		calories	joules	lit-atms
1 calorie	=	1	4.184	0.04129
1 joule	=	0.2390	1	0.00987
1 lit-atm	=	24.22	101.3	1

1. Suppose we had chosen to base our temperature scale *not* on the triple point but, rather, on the normal freezing point of water. Denoting by P_B and P_F the pressures actually measured by a constant-volume gas thermometer at the boiling temperature (T_B) and the freezing temperature (T_F) of water, we find that

$$\lim_{P \to 0} [P_B/P_F]_V = 1.3661.$$

If we wish to maintain a 100° interval between the freezing and boiling points of water, what number should we assign as the temperature of freezing water?

2. (a) For the two lines graphed in Fig. 4, Eberhardt gives the following equation:

$$w = \frac{Mg\,\Delta h}{2}\left(1 \mp \frac{\delta M}{M}\right),$$

where Δh represents the difference in height between states I and II and the minus and plus signs refer to the work output and work input respectively. Keeping in view the detailed calculations on pp. 9–12, you should find it easy to derive this equation. Do so!

(b) The above equation permits calculation of the finite difference between work output and work input when the changes I $\to$ II and II $\to$ I are conducted with some particular finite δM. Obtain an expression for this difference, and relate it to the distribution of weights shown, for example, in the columns labeled (d) and (h) in Figs. 3 and 5 respectively.

3. (a) Between room temperature and its melting point at 327°C, lead has a specific heat of ~0.033 cal/gm-°C. A bullet at an initial temperature of 20°C is stopped by impact with an unyielding target. Assuming no loss of heat from the bullet to the target, how fast must the bullet be traveling if it is just brought to its melting point by the impact?

(b) A powerful steel spring, weighing 55.8 gm, is squeezed 2 cm shorter by a force averaging 100 kgf (=980 newtons). Given that ~21 kcal are liberated by the solution of 1 gm-atom of iron in dilute hydrochloric acid, what would have to be the percentage accuracy of calorimetric measurements designed to detect the increase in the internal energy of the spring as a difference in the heats of solution of the stressed spring and an otherwise identical unstressed spring?

4. (a) On the occasion of his honeymoon excursion to Switzerland, J. P. Joule is said to have rejoiced in the discovery that the water at the base of a waterfall is perceptibly warmer than at its top. Assuming that he carried a thermometer sensitive enough to show a temperature difference of 0.1°C, what was the minimum height of the waterfall he visited?

(b) A corpulent man, with mass 100 kgm, seeks to compensate for putting several lumps of sugar in his breakfast coffee by climbing the almost vertical rocks alongside the waterfall of part (a). His (typical animal) metabolism permits the conversion into work of approximately $\frac{1}{4}$ the heat of combustion of his food. Given that the heat of combustion of cane sugar is 1350 cal/gm, how many grams of sugar suffice to furnish the energy for his climb?

5. (a) During expansion against a constant external pressure of 1 atm, a gas absorbs 50 cal while its volume increases from 1 to 10 liters. What is the net change in its internal energy consequent to this expansion?

(b) If one mole of ideal monatomic gas undergoes the expansion described in (a), by how many degrees will its temperature be changed?

6. (After Kauzmann) The minimum "escape velocity" that permits a body to free itself from the earth's gravitational field is an easily calculable quantity of the order of 25,000 miles per hour. Consider the contrary case in which an object plunges toward the earth with a velocity of 25,000 mph. Per gram of object, calculate the energy released in a head-on collision with the earth, and compare your result with the 3.5×10^3 kcal representing the energy released in the detonation of 1 gram of TNT. (For a fictional parable involving related collisional effects, see *The Black Cloud* by Fred Hoyle.)

7. (a) If, in addition to $P\,dV$ work, some electrical work (w^*) is delivered by a system undergoing change, show that equations (3) and (5) must be rewritten as respectively:

$$q_V = \Delta E + w^*, \qquad q_P = \Delta H + w^*.$$

(b) We symbolize by ΔE_P the change of internal energy in a constant-pressure process for which the heat transfer is q_P; and by ΔH_V the change of enthalpy in a constant-volume process for which the heat transfer is q_V. Starting from equation (6), show that

$$\Delta E_P = q_P - P\,\Delta V \qquad \text{and} \qquad \Delta H_V = q_V + V\,\Delta P.$$

8. (a) In discussing the characteristics of energy as a function of state, we have noted that (i) around a closed cycle $\Delta E_{ABA} = 0$, and (ii) for any two paths between the same end states $\Delta E_{I} = \Delta E_{II}$. Show that, for functions of state generally, the characteristics (i) and (ii) are redundant in that each implies the other.

(b) On p. 35 we derive our first expression of Hess's law by a development beginning from characteristic (ii). Now derive the expression $\Delta H_1 = \Delta H_2 + \Delta H_3 + \Delta H_4$ by setting out from characteristic (i), and a redrawn Fig. 14.

9. At 25°C and constant pressure, 326.7 kcal are released in the combustion of one mole of ethanol according to the equation

$$C_2H_5OH\ (\ell) + 3O_2\ (g) = 2CO_2\ (g) + 3H_2O\ (\ell).$$

(a) For the reaction, as written, what is the value of ΔH? of ΔE?

(b) At 25°C, 10.5 kcal are required to vaporize a mole of water. If in the above reaction the water had been obtained not as liquid but as gas, what then would have been the value of ΔH? of ΔE?

(c) At 25°C and constant pressure, ΔH_f for H_2O (ℓ) is -68.3 kcal/mole, and ΔH_f for CO_2 (g) is -94.1 kcal/mole. Calculate ΔH_f for C_2H_5OH (ℓ) under these conditions.

(d) For C_2H_5OH (g) the value of ΔH_f is -56.2 kcal at 25°C and constant pressure. For the vaporization of 1 mole of ethanol under these conditions, what is the value of ΔH? of ΔE?

10. (a) A small Dewar vessel, containing a 10-ohm resistor and a sensitive thermometer, is used as a calorimeter. When 0.01 mole of metallic zinc is added to excess dilute aqueous HCl in this calorimeter, a temperature rise of 2.429°K is swiftly produced by the reaction: $Zn\ (s) + 2H^+\ (aq) = Zn^{++}\ (aq) + H_2\ (g)$. The calorimeter and its contents are then permitted to cool slowly back to their original temperature. At this point a 12.00-volt difference of potential is applied for 2.00 minutes to the 10.00-ohm resistor, and the total temperature rise so produced is found to be 2.757°K. Indicating any special assumptions used in obtaining your answer, calculate ΔH, in kcal/mole Zn, for the reaction written above.

(b) In a similar determination, the addition of 0.01 mole of ZnO to the same volume of equally dilute aqueous HCl yields $\Delta H = -21.5$ kcal/mole ZnO for the reaction $ZnO\ (s) + 2H^+\ (aq) = Zn^{++}\ (aq) + H_2O\ (\ell)$. Given that for liquid H_2O, $\Delta H_f = -68.3$ kcal/mole, calculate the heat of formation of ZnO —that is, ΔH for the reaction $Zn\ (s) + \frac{1}{2}O_2\ (g) = ZnO\ (s)$.

(c) The calculation in part (b) involves a major implicit assumption—well justified in the present case by good agreement of the calculated value with that measured directly. What is that assumption?

11. There exist three isomeric pentanes with molecular formula C_5H_{12}: normal pentane, $CH_3CH_2CH_2CH_2CH_3$; isopentane, $CH_3CH_2CH(CH_3)_2$, and neopentane, $C(CH_3)_4$. Each of these compounds contains 4 C—C bonds and 12 C—H bonds, and each readily enters the combustion reaction:

$$C_5H_{12}\ (g) + 8O_2\ (g) = 5CO_2\ (g) + 6H_2O\ (\ell).$$

(a) The molar heat of combustion of normal pentane, in the indicated reaction, is -845.2 kcal. Taking -94.1 and -68.3 kcal as the molar heats of formation of CO_2 (g) and H_2O (ℓ) respectively, determine the heat of formation of normal pentane.

(b) The molar heats of combustion of isopentane and neopentane are —843.3 and —840.5 kcal respectively. By inspection of the result obtained in (a), write down the heats of formation of isopentane and neopentane. From the viewpoint of bond energies, should all three heats of formation be the same?

(c) Using the bond-enthalpy data given in Appendix IV, and taking +171 kcal as the molar heat of sublimation of graphite and +104 kcal as the molar heat of dissociation of H_2 (g), calculate the expected heat of formation of a pentane.

(d) What is the percentage difference between the theoretical value obtained in (c) and the average of the three basically empirical values obtained by Hess-law analysis in (a) and (b)?

12. Imagine an ordinary room filled with an ideal gas. Writing *per mole* of gas that $dE = C_V\,dT$, show that—over the temperature range in which C_V is constant—the value of E for the gas *in the room* is wholly independent of temperature. [See R. Emden, *Nature* **141,** 908 (1938).]

13. (a) Derive the equation: $d(\Delta H) = \Delta C_P\,dT$.

(b) Insert in this equation the expression for ΔC_P given in the footnote on p. 47 and, taking the *indefinite* integrals, show that the enthalpy of reaction (ΔH_T) at any temperature (T) can be written as:

$$\Delta H_T = \Delta H_0 + (\Delta a)T + \frac{\Delta b}{2}T^2 - \frac{\Delta c}{T},$$

where ΔH_0 is the constant of integration.

(c) How might you try to establish the actual magnitude of ΔH_0?

(d) Subject to an arbitrary assumption about the term Δc, a plausibly simple interpretation of the constant ΔH_0 becomes possble. Identify the assumption and state the interpretation.

(e) Your interpretation in (d) may seem to be invalidated by the fact that expressions for ΔC_P as $f(T)$ are usually applicable only within some such temperature range as 298°–2000°K. To what extent does this fact invalidate the equation given in (b) and/or the interpretation you have given under (d)?

14. At 100°C the heat of vaporization of water is 9.7 kcal/mole, and the heat capacities of liquid and vapor are respectively ~18 and 8 cal/mole-°C.

(a) Use Kirchhoff's equation to estimate the heat of vaporization of water at 25°C.

(b) Although the result obtained in (a) is in fair agreement with the correct value (~10.5 kcal), this use of Kirchhoff's equation is fundamentally invalid. Why?

15. (a) Relative to the strongly exothermic reaction $X + Y = Z$, the text's equation $(T_f - T_i) = -\Delta\overline{E}_i/(C_V)_Z$ is obtained (on p. 49) on the assumption that $(C_V)_Z$ remains the same throughout an immense range of temperatures. This potentially vulnerable assumption is easily avoided by expressing $(C_V)_Z$ as a power series, in the manner indicated in the footnote on p. 40. What expression does one then obtain for the relation of the temperature change $(T_f - T_i)$ to the energy change $\Delta\overline{E}_i$?

(b) Suppose one sought a value for the energy change ($\Delta\overline{E}_f$) that would characterize the indicated reaction were it to run its entire course at the peak temperature (T_f) attained in the adiabatic explosion in a constant-volume bomb. Assuming effective constancy of known heat capacities $(C_V)_X$ and $(C_V)_Y$, $\Delta\overline{E}_f$ is readily determinable from the observed temperature rise. Derive the relevant equation, which is

$$T_f - T_i = \frac{-\Delta\overline{E}_f}{(C_V)_X + (C_V)_Y}.$$

16. (a) A stoichiometric mixture of gaseous hydrogen with air (taken as 80% N_2 and 20% O_2) is fed through the nozzle of a torch. Assume an initial temperature of 298°K and a pressure constant throughout at 1 atm; assume further that there is no loss of heat from the flame to the surroundings. For H_2O (g) take C_P as 9 cal/mole-°C, for N_2 (g) take C_P as 7 cal/mole-°C, and use $\Delta H_{298} =$ —58 kcal for the overall reaction

$$H_2\ (g) + \tfrac{1}{2}O_2\ (g) = H_2O\ (g).$$

Noting that both H_2O (g) and N_2 (g) are present after the combustion, calculate the peak temperature achievable.

(b) To fuse a patch over a rent in the high-melting envelope of my spaceship, I use a torch that sprays the metal with a stream of atomic hydrogen, produced by passing H_2 (g) through an electric field. The reaction produced as the atomic hydrogen hits the metal is

$$2H\ (g) = H_2\ (g), \qquad \text{with } \Delta H_i = -104 \text{ kcal}.$$

Assuming constant (low) pressure, 100% efficiency in the generation of H (g), and no heat loss to the surroundings, and taking as C_P for H_2 (g) a value of 7 cal/mole-°C, calculate the maximum temperature rise I can produce with my torch.

(c) The actually attainable peak temperature is far less than that calculated in (b). To what limitation(s) *in principle* would you attribute the discrepancy of calculated and attained temperatures?

17. (a) Sometimes it is easier to measure the peak pressure (P_f) than to measure the peak temperature (T_f) produced when, in a bomb of volume V, n moles of Z are formed by the strongly exothermic reaction X (g) + Y (g) = Z (g). What is the relation of T_f to P_f?

(b) The peak temperature reached in a constant-volume explosion is generally *much* higher than the peak temperature reached when the same reaction is carried out in a constant-pressure flame. Assuming that heat losses can be ignored in both cases, determine which of the following effects is chiefly responsible for the difference of peak temperatures: (i) difference of ΔE_i from ΔH_i; (ii) difference of $(C_V)_Z$ from $(C_P)_Z$; (iii) difference of pressure-volume work performed.

(c) At 298°K the molar heats of combustion of acetylene (C_2H_2) and ethane (C_2H_6) are:

$$C_2H_2\ (g) + \tfrac{5}{2}O_2\ (g) = 2CO_2\ (g) + H_2O\ (\ell), \qquad \Delta H = -311 \text{ kcal};$$
$$C_2H_6\ (g) + \tfrac{7}{2}O_2\ (g) = 2CO_2\ (g) + 3H_2O\ (\ell), \qquad \Delta H = -373 \text{ kcal}.$$

When a stoichiometric mixture of acetylene and oxygen is exploded in a sealed bomb, the temperature rises considerably *higher* than when a stoichiometric mixture of ethane and oxygen is exploded under exactly matching conditions. Explain!

18. A stoichiometric mixture of methane, CH_4, and air is exploded in a sealed bomb. Assuming no heat loss to the walls of the bomb, and taking air as 20% O_2 and 80% N_2, the initial temperature as 25°C, and the initial pressure as 1 atm, take as the values of C_V for N_2 (g) 5 cal/mole-°C, for H_2O (g) 7 cal/mole-°C, and for CO_2 (g) 7 cal/mole-°C. The peak pressure attained in the explosion is found to be 9.5 atm. Keeping in view the relation derived in 17(a):

(a) Calculate the difference between ΔH_{298} and ΔE_{298} for the reaction:

$$CH_4\ (g) + 2O_2\ (g) = 2H_2O\ (g) + CO_2\ (g).$$

(b) How large is ΔH_{298} for the combustion of CH_4 in this reaction?

(c) What considerations might explain the rather wide divergence of the measured result from the correct value for ΔH_{298}? Might one hope to do better by running the reaction with pure oxygen rather than with air?

19. (After Waser) The vernier rockets used to maneuver space vehicles may be fueled with an aqueous solution that is 75% by weight H_2O_2. When the rocket is to be actuated, this solution is forced through a catalyst that brings about rapid completion of the reaction:

$$H_2O_2\ (\ell) = H_2O\ (\ell) + \tfrac{1}{2}O_2\ (g), \quad \Delta H_{298} = -23.4\ \text{kcal}.$$

If the solution enters the catalyst chamber at 298°K, at what temperature will the reaction products leave that chamber? What auxiliary simplifying assumptions have you made?

As values of C_P use for H_2O (ℓ), 18 cal/mole-°K; for H_2O (g), 9 cal/mole-°K; for O_2 (g), 7 cal/mole-°K. As the molar heat of vaporization of liquid water use $\Delta H_{\text{vap}} = 9.7$ kcal/mole, and assume as molecular weights 18 and 34 for H_2O and H_2O_2 respectively.

20. While doing work against a constant external pressure of 2 atm, two moles of an ideal gas expand from 5 to 15 liters at a constant temperature of 300°K.

(a) For the gas undergoing this expansion, what is the value of ΔE? of ΔH?

(b) Calculate the work done and the heat absorbed by the gas.

(c) Comment on the possibility of basing on this expansion an engine for high-efficiency continuous conversion of heat into work.

21. (a) According to equation (14), the work done in the *adiabatic* expansion of an ideal gas is expressible as $w = -nC_V(T_2 - T_1)$ for a gas of constant heat capacity. Show that this equation, taken together with the ideal gas law, $PV = nRT$, at once gives the following expression for the work done during the expansion:

$$w = \frac{P_1V_1 - P_2V_2}{\gamma - 1}.$$

(b) Two moles of an ideal monatomic gas ($C_V = 3$ cal/mole-°K) expand irreversibly from an initial pressure of 10 atm, against a constant external pressure of 1 atm, until the temperature drops from the initial value of 325°K to a final value of 275°K. How much work is done, and what is the final volume?

22. (a) On a plot of P *vs* V, the equation PV^γ = constant describes the line representing the reversible adiabatic expansion of an ideal gas, while the equation PV = constant describes the line representing the corresponding isothermal expansion. By deriving expressions for the slopes (dP/dV) of the two lines, show that at any point the adiabatic line is γ times steeper than the isotherm running through the same point.

(b) Setting out from the ideal gas law and either equation (15) or (16), show that the reversible adiabatic expansion of an ideal gas is also describable by a third equation: namely,

$$\frac{P_1}{T_1^{C_P/R}} = \frac{P_2}{T_2^{C_P/R}}.$$

23. As three illustrations of the usefulness of the relation stated in part (b) of the preceding problem, consider:

(a) A fire syringe is a dead-end cylinder with a tightly fitting piston to the inner face of which is attached a bit of tinder. Using a syringe of 1 in^2 cross section, a 147-lb man applies all his weight to the compression of air originally at a temperature of 300°K and a pressure of 1 atm (i.e., 14.7 lb/in^2). Assuming the compression reversible and adiabatic, and assuming further that air is an ideal gas with $C_P = \frac{7}{2}R$, determine the maximum kindling temperature of tinder that must ignite by the end of the compression stroke.

If you know anything of the distinctive feature of the diesel engine, indicate the relevance of this kind of computation to the design of diesel engines.

(b) Liquid air boils at ~ -190°C. Assume air to be an ideal gas with $C_P = \frac{7}{2}R$. What must be the initial pressure of air at 0°C if its reversible adiabatic expansion to a final pressure of 1 atm is barely to cool it to its boiling point?

Air is a *non*ideal gas, between the molecules of which there are small but finite attractive forces. Will this nonideality make the actual cooling greater or less than that calculated above?

(c) In the expansion nozzle of a jet engine, thrust is generated by a change reasonably well approximated as the reversible adiabatic expansion of an ideal gas. Assume that the nozzle is fed from a combustion chamber producing gas (with molecular weight 18 and $C_P = \frac{9}{2}R$) at a temperature of 2000°K and a pressure of 30 atm, and that at its exit end the nozzle delivers gas effectively at 1 atm pressure. Calculate (i) the temperature of the exit gas, and (ii) the velocity to which it has been accelerated.

24. On hills and mountains the temperature is generally less than that at sea level; and six miles up, around our airliner, the temperature may sink more than 100°F below that at sea level. Now large masses of air are continually rising and falling through the atmosphere and, as the pressure incumbent upon them changes, they expand or contract. Due to their large mass, they move slowly enough that we can regard the expansions and contractions as *reversible*, and all kinetic-energy effects as negligible. And given their large mass and the small conductivity of air, we may regard the expansions and contractions as *adiabatic*. In appraising the effect of these reversible adiabatic alterations within the atmosphere, we treat air as an ideal gas with average molecular weight M = 29, and heat capacity $C_P = \frac{7}{2}R$. And we seek now to evaluate the derivative dT/dh,

which expresses the variation of atmospheric temperature (T) with altitude (h) above sea level.

(a) For a slab of atmosphere under pressure P at altitude h, the pressure difference (dP) between the top and bottom of a slab of thickness dh arises simply from the mass of the slab itself. Show that $dP/P = -(Mg/RT)\,dh$, where g symbolizes the gravitational acceleration.

(b) The equation appearing in problem 22(b) applies to reversible adiabatic expansions and compressions of ideal gases. From that equation obtain the relation: $dP/P = C_P\,dT/RT$.

(c) Substituting from (b) in (a), derive a simple expression for the derivative dT/dh. Making appropriate numerical substitutions, with due attention to units, obtain for this "lapse rate" a value of the order of $-10°$K/kilometer.

(d) The lapse rate calculated in (c) is in fair agreement with the observed average variation of temperature with altitude, but is somewhat too large. Would correction for the thermal effect of the condensation of water droplets from humid air act to increase or decrease the divergence, and why?

25. (a) On p. 59 the conclusion that $W_{\text{i}} > W_{\text{ii}}$ is based on nothing more than the superficial appearance of admittedly distorted graphical representations. Show how this same conclusion can be reached analytically, on the basis of the familiar equations for the work done in reversible isothermal and adiabatic expansions of an ideal gas.

(b) When (on p. 62) we first considered how materials other than ideal gases might be carried through the Carnot cycle, we tacitly excluded the possibilities that an engine might deliver a work output ($+W$) by (i) absorbing heat $+Q_L$ at temperature T_L and rejecting a smaller amount of heat $-Q_H$ at temperature T_H, or by (ii) absorbing heat $+Q_L$ at temperature T_L and absorbing heat $+Q_H$ at temperature T_H. Neither possibility conflicts with the first principle, and neither can be excluded on the strength of Carnot's theorem—in the derivation of which such possibilities were already excluded. How then can you rule out possibilities (i) and (ii)?

(c) On p. 64 we show that a *reductio ad absurdum* arises inescapably from the assumption that, of two ideal heat engines operating over the same temperature range, one is less efficient than the other. Explain why, if a real engine and an ideal engine operate over the same temperature range, the same *reductio ad absurdum* DOES NOT arise if the real engine proves (as it always does) less efficient than the ideal engine, though it WOULD arise if the real engine were to prove (as it never does) more efficient than the ideal engine. *Hint:* what characteristic connoted by "ideal" is certainly absent from the "real."

26. (a) For a refrigerator, the ratio $Q_L/-W$ serves as a factor of merit in much the same sense that W/Q_H is a factor of merit for a heat engine, and $-Q_H/-W$ is the corresponding factor of merit for a heat pump. Explain!

(b) Starting from equation (20), show that for an ideal refrigerator:

$$\frac{Q_L}{-W} = \frac{T_L}{-(T_H - T_L)}.$$

27. (a) A reversible engine 1 operates (as engines generally do) between some high temperature (T_H) and an exhaust temperature (T_M) which is the ambient

temperature of the surroundings. We may think to make better use of the high-temperature heat input by replacing engine 1 with reversible engine 2. This operates between T_H and an exhaust temperature (T_L) held far below the temperature of the surroundings by the continuous operation of a refrigerator, which delivers its heat output at temperature T_M as shown in the accompanying figure. Show that engine 2 is more efficient than engine 1 in the conversion of heat input into work output.

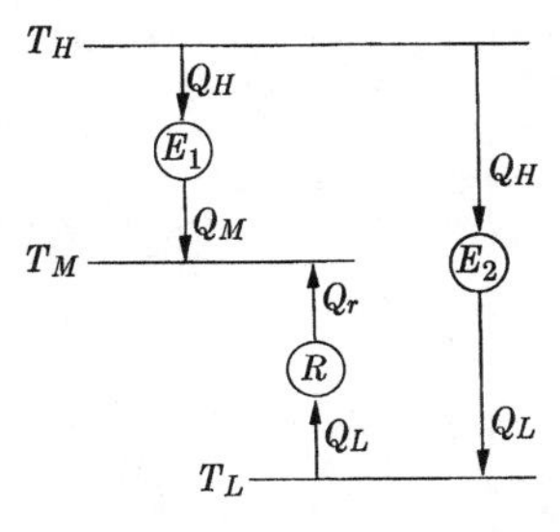

PROBLEM 27

(b) The answer just obtained surely doesn't tell the whole story, because part of the work output (W_2) delivered by engine 2 will be required to operate the refrigerator that removes the heat delivered by the engine at temperature T_L. Symbolizing by $-W_r$ the work input needed to make a reversible refrigerator pick up heat Q_L at temperature T_L, show that

$$-W_r = \frac{-(T_M - T_L)}{T_L} Q_L.$$

(c) From the efficiency formula for engine 2, obtain a simple expression for Q_L in terms Q_H; substituting this expression in the last equation above, show that $-W_r$ is related to the high-temperature heat input to engine 2 by the equation:

$$-W_r = \frac{-(T_M - T_L)}{T_H} Q_H.$$

(d) Obtain a simple expression for the difference ($W_2 - W_r$), which represents the *net* work output of the ensemble of engine 2 plus refrigerator. Comment on the relation of this result to the efficiency formula for engine 1.

28. (a) Into a small closed room is brought a brand new electric refrigerator, which is then plugged in with its door wide open. What happens to the temperature of the room?

(b) An ice cube weighs about 36 gm (=2 moles of water). In making one ice cube at 0°C from water at 0°C, how many kilowatt-hours of energy will be used by an ideal Carnot refrigerator, standing in a room at 20°C, if for the conversion of water to ice $\Delta H = -1440$ cal/mole at 0°C, and 1 kwh = 3.6×10^6 joules?

29. In generalizing the Carnot cycle to all materials, we assumed implicitly that, for any given material, two adiabatic lines never intersect. Demonstrate the validity of this assumption by showing that its falsity would make possible a cyclic change with the 100% conversion of heat into work that we never observe

in such a change. [This customary "derivation" is warmly criticized by J. Kestin, *Am. J. Phys.* **29**, 329 (1961).]

30. In the manner that one indicates a change of state by a suitably placed line segment on a P *vs* V plot, one can also indicate the change on a plot of T *vs* S.

(a) On the P-V plot the area beneath the line segment represents the pressure-volume work for the change in question. What is represented by the similarly placed area on the T-S plot?

(b) On a T-S plot, draw a line to represent a reversible adiabatic change.

(c) On a T-S plot, diagram the Carnot cycle for an ideal gas.

(d) Show how the Carnot engine-efficiency formula can be read off directly from the diagram you have drawn in (c).

[On T-S plots more generally, see T. S. Breck, *J. Chem. Educ.* **40**, 353 (1963).]

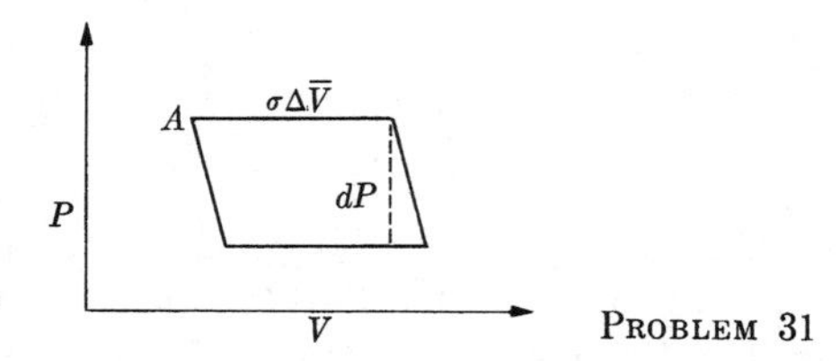

PROBLEM 31

31. Imagine a *mixture* of liquid and its vapor, carried together through the Carnot cycle shown in the sketch. Starting at A, a minute smidgen (σ) of liquid is vaporized isothermally, with a consequent change of $\sigma\,\Delta\overline{V}$ in the volume of the system. There follows an infinitesimal adiabatic expansion of the mixture, with temperature and pressure dropping by dT and dP respectively. Then come successive isothermal and adiabatic compressions that at last return the system to the original point A.

(a) Why do the isotherms here lie horizontally?

(b) Approximating the complete cycle as a parallelogram, show that $W = \sigma\,\Delta\overline{V}\,dP$ and $Q_H = \sigma\,\Delta\overline{H}$, where $\Delta\overline{H}$ is the molar heat of vaporization.

(c) Applying the Carnot efficiency formula to this cycle operating over the infinitesimal temperature range dT, obtain an expression for dP/dT, i.e., for the temperature dependence of the vapor pressure of the liquid. Comment on the relation of your result to equation (45) on p. 107.

32. In the style of pp. 72–73, demonstrate that (i) No manner of conducting an isothermal change recovers from it more work than is recoverable when the change proceeds reversibly; and (ii) When a given change of state can proceed along different isothermal routes, the maximum work recoverable along any route is the *same*.

33. When a rubber band is stretched, the long polymeric molecules (which fall in random coils in unstretched rubber) are drawn into a more highly-ordered array of nearly parallel linear molecules. But in this stretching no chemical bonds are broken, or even significantly distorted, and for the *isothermal* stretching of rubber $\Delta E = 0$ to a good first approximation. Making constant reference

to the cognate case of isothermal expansion and compression of an ideal gas, analyze the reversible and irreversible stretching and relaxation of rubber. Note specifically the transfers of heat and work that occur, as well as the consequent changes in the system's entropy and capacity to deliver work.

34. (a) A closed cylinder is fitted with three diaphragms, as shown in the accompanying figure.

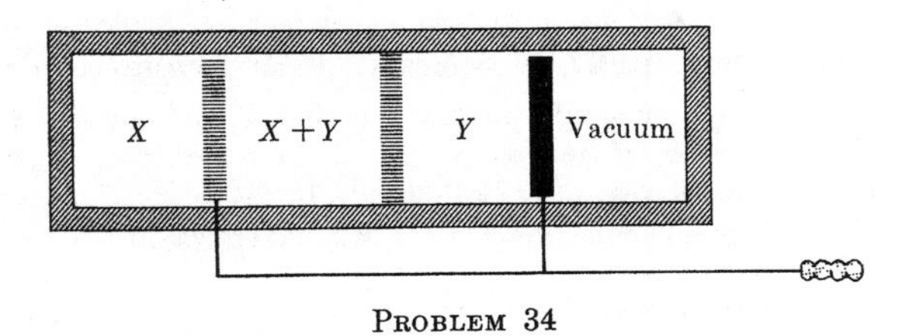

PROBLEM 34

The cylinder is divided in half by a fixed diaphragm selectively permeable to gas Y. The other two diaphragms, of which the one on the left is selectively permeable to gas X and the one on the right is totally impermeable, move together as a linked unit. How much work and heat will be involved in the (i) mixing and (ii) separation of ideal gases that this device makes it possible to conduct reversibly?

(b) A gas X exerts a pressure P_X in some particular volume V; at the same temperature another gas Y exerts a pressure P_Y in *another* numerically equal volume V. The two gases are brought together isothermally so that they occupy the *same* volume V and exert a total pressure $(P_X + P_Y)$. What is the value of ΔS for this change, and how is it reconciled with the randomization that apparently occurs when the gases are mixed?

35. Imagine an isolated system—adiabatic and constant in total volume—in which there initially exists some inequality of temperature or pressure. We wish to show that dissipation of this inhomogeneity is a spontaneous process, easily recognizable as such because it proceeds with a net increase of entropy.

(a) Let the system consist of 1 mole of ideal gas A, occupying a volume V_A, separated by a movable but impermeable piston from 1 mole of ideal gas B, occupying a different volume V_B at the same temperature (T). When the piston has moved until the two gases, still at temperature T, stand at equal pressures in equal volumes, show that the total entropy change of the two gases is given by the expression:

$$\Delta S = R \ln \frac{(V_A + V_B)^2}{4\,V_A V_B}.$$

(b) By appropriate rearrangement of the last formula, show that it always involves the logarithm of a number greater than 1, so that the equalization of pressures must always proceed with $\Delta S > 0$.

(c) Consider now a system consisting of two moles of metal, of which 1 mole stands initially at a uniform temperature T_A while the other mole stands initially at a different uniform temperature T_B. Assuming that the metal's heat capacity (C_P) is constant over this range of temperature, the transfer of heat will ultimately equalize the temperatures at a value of $(T_A + T_B)/2$. Show that the change of entropy accompanying this equalization of temperatures is given by the relation:

$$\Delta S = C_P \ln \frac{(T_A + T_B)^2}{4\,T_A T_B}.$$

(d) Will the argument of part (b) be applicable here, to demonstrate that the equalization of temperatures is accompanied by an increase of entropy?

36. The approach to reversibility in heat transfers is well illustrated by Dugdale, through consideration of how one may cool from 200° to 100°K a block of metal with a constant heat capacity of 1000 cal/°K.

(a) Show that in this cooling the block itself undergoes an entropy change of —693 cal/°K.

(b) Suppose that the cooling is brought about simply by plunging the block into a large bath at 100°K. Show that the entropy change of the block and the bath together is then +307 cal/°K.

(c) Suppose that the block is cooled in two stages: first by immersion in a large bath at 150°K, then in a large bath at 100°K. Show that the entropy change of the block and the baths together is then +140 cal/°K.

(d) Suppose that the block is cooled in four stages, using baths at 175°, 150°, 125°, and 100°K. Show that the entropy change of the block and the baths together is then +67 cal/°K.

(e) Relate these results to the analysis of reversibility in Chapter 1. Describe the limiting process in which the block is cooled reversibly, and indicate the total entropy change of the block and the baths consequent to this process.

37. (a) If increase in entropy (and disorder or "spread") is the *sine qua non* of a spontaneous process, how is it possible for a liquid, once vaporized, ever to recondense; or for a solid, once fused, ever to recrystallize spontaneously?

(b) Let a mole of liquid at its normal freezing point (T_F) be transformed at a finite rate into a mole of solid at the same temperature. For the total consequent entropy change of this system *and* its surroundings, the appropriate expression is

$$\Delta S = \Delta \overline{H}_{\text{fus}} \left[\frac{T_F - T}{T T_F} \right],$$

where $\Delta \overline{H}_{\text{fus}}$ is the molar heat of fusion of the solid and T is the (constant) temperature of the surroundings. Derive this expression, and indicate why *necessarily* $\Delta S > 0$ if the solidification proceeds at a finite rate.

38. (a) Let n_A moles of ideal gas A occupy volume V_A, while n_B moles of ideal gas B occupy a separate volume V_B at the same temperature and pressure. Show that when the two volumes are connected, and the gases are allowed to mix isothermally, their total entropy change is:

$$\Delta S = -nR \sum X_i \ln X_i,$$

where $n = n_A + n_B$ and the X_i's symbolize the mole fractions of A and B in the resultant mixture. Show further that always $\Delta S > 0$.

(b) Assuming all gases ideal, calculate the work invested, the heat dissipated, and the change of entropy when 100 liters of air at 1 atm pressure and 298°K are separated isothermally and reversibly into 79 liters of nitrogen and 21 liters of oxygen measured at 1 atm pressure and 298°K.

39. At 0°C the heat of fusion of ice is 1440 cal/mole and the C_P values are 18 cal/mole-°C for water and 9 cal/mole-°C for ice.

(a) One mole of water, set out on a cold night, supercools to the air temperature of −10°C. By analyzing a series of steps through which the end result is attained reversibly, calculate the entropy change of the system when the water at −10°C freezes to ice at the same temperature.

(b) How do you account for the negative value of the entropy change calculated in part (a) for what is obviously a spontaneous process? Calculate the net overall entropy increase for the process.

40. (After Cole and Coles) (a) One mole of nitrogen, standing initially at 1 atm pressure and 25°C, is compressed reversibly and adiabatically to a final pressure of 20 atm. Taking $C_P = 6.95$ cal/mole-°K, show that at the end of the compression the gas stands at 430°C in a volume of 2.88 liters. What is ΔS for this change?

(b) Consider a second route connecting the same initial and final states. Let the nitrogen first be compressed isothermally, at 25°C, to a final pressure of 20 atm; then let the gas be heated at constant pressure to a final temperature of 430°C. Calculate ΔS for each of the two steps, and add to find ΔS for the complete route.

(c) Consider a third route connecting the same initial and final states. Let the nitrogen first be compressed isothermally, at 25°C, to a final volume of 2.88 liters; then let the gas be heated at constant volume to a final temperature of 430°C. Calculate ΔS for each of the two steps, and add to find ΔS for the complete route.

41. At its normal boiling point of 373°K water has a heat of vaporization of 545 cal/gm. Taking 18 as the molecular weight of water, and regarding its vapor as an ideal gas *far* more voluminous than the corresponding liquid, calculate the values of q, w, ΔH, ΔE, ΔS, ΔA, and ΔG for the reversible vaporization of 1 mole of water at 373°K.

42. (a) Setting out from equation (41), $dE = T\,dS - P\,dV$, derive the relations:

$$dH = T\,dS + V\,dP \qquad \text{and} \qquad dA = -S\,dT - P\,dV.$$

(b) Setting out from equation (44), $d(\Delta G) = \Delta V\,dP - \Delta S\,dT$, derive the Clapeyron equation in the form $dP/dT = \Delta H/T\,\Delta V$.

(c) Derive the following two additional Gibbs-Helmholtz relations:

$$\frac{d}{dT}\left[\frac{\Delta A}{T}\right]_V = -\frac{\Delta E}{T^2} \qquad \text{and} \qquad \left[\frac{d(G/T)}{d(1/T)}\right]_P = H.$$

43. Under a pressure of 1 atm, rhombic sulfur is stable below 368.5°K, monoclinic sulfur is stable above 368.5°K, and the two allotropic species stand in

equilibrium at 368.5°K. Consider now the reaction:

$$S(\text{rhombic}) \rightarrow S(\text{monoclinic}).$$

(a) What must be the *sign* of $\Delta\bar{S}$ in the vicinity of 368.5°K?

(b) Given that $\Delta\bar{H} = 96$ cal, determine $\Delta\bar{S}$ for the indicated reaction.

(c) Given that $\Delta\bar{V} \cong 0.4$ ml for this reaction, determine the temperature at which the two species would stand in equilibrium under a pressure of 100 atm.

(d) Treating $\Delta\bar{H}$ and $\Delta\bar{V}$ as constants, we can determine the effect of a moderate change of pressure on a transition temperature (or melting point) by using dT/dP just as it is given by equation (45). But in the corresponding determination of the effect of a moderate change of pressure on a boiling point, we are better advised to recast equation (45) as $d \ln T/dP = \Delta\bar{V}/\Delta\bar{H}$. Why?

44. (a) Water has a heat of vaporization of 9.7 kcal/mole near its normal boiling point of 100°C. At what temperature will water boil atop Pike's Peak under the pressure of ~0.6 atm there prevailing at 14,100 ft?

(b) For any liquid that obeys Trouton's rule, and boiling at temperature T_B under a pressure of 1 atm, derive the relation

$$\ln P = 10.5\left(1 - \frac{T_B}{T}\right)$$

as an expression for the variation of vapor pressure (P) with temperature (T).

(c) The relation derived in (b) facilitates rapid order-of-magnitude estimates. Appraise its reliability in the following two calculations. (i) At what temperature can cymene, with $T_B = 177$°C, be vacuum distilled at a pressure of 10 mm mercury? The temperature actually measured is 57°C. (ii) Mercury, with $T_B = 357$° C, should exert what vapor pressure at 18°C? The pressure actually measured is 0.001 mm mercury.

45. (a) Solid benzene has a vapor pressure of 1.0 mm at −38.5°C and 24.5 mm at 0°C. Liquid benzene has a vapor pressure of 26.7 mm at 0°C and 100 mm at 26.1°C. By suitable graphical treatment of these data determine (i) the heat of vaporization of liquid benzene, (ii) the heat of sublimation of solid benzene, (iii) the heat of fusion of benzene, and (iv) the melting point of benzene.

(b) Moderate increases in applied pressure increase the melting point of benzene by 0.03°C/atm. Given that at the melting point the molar volumes of solid and liquid benzene are respectively 76.5 cm^3/mole and 87.6 cm^3/mole, and using the melting point calculated in (a), determine the heat of fusion of benzene.

(c) Considering the data and the methods involved, how do you appraise the agreement of the results obtained in part (b) and in (iii) of part (a)?

46. On the basis of earlier work by Jackson, in the temperature range 150–300°C, McMorris and Yost report [*J. Am. Chem. Soc.* **53**, 2625 (1931)] that in the reaction $2\,CuBr_2$ (s) = $2\,CuBr$ (s) + Br_2 (g) the equilibrium pressure of Br_2 is given by the formula: $\log P(\text{mm}) = -4921.2/T + 11.6682$. Other available data relevant to the same reaction are:

	Br_2 (g, 1 atm)	$CuBr$ (s)	$CuBr_2$ (s)
ΔH^0_{f298} (kcal/mole)	+7.34	−25.1	−33.2
S^0_{298} (cal/mole-°K)	+58.64	+21.9	

The NBS Circular 500 gives no value for S^0_{298} of $CuBr_2$ (s), and we proceed now to obtain the value used in illustrative Example 2 (p. 100).

(a) Derive the equation: $\log P(\text{atm}) = -4921.2/T + 8.7874$.

(b) Derive from the equation in (a) a figure for ΔH^0 in the indicated reaction.

(c) Derive from the tabulated data another, independent figure for ΔH^0.

(d) What basis have you now for the assertion that both ΔH^0 and ΔS^0 are substantially constant over the temperature range 25–300°K?

(e) From the equation in (a), calculate the temperature at which the equilibrium pressure of Br_2 is 1 atm.

(f) What is the value of ΔS^0 in the indicated reaction?

(g) What is the value of S^0_{298} for $CuBr_2$ (s)?

47. Spontaneous freezing of supercooled water at −10°C proceeds highly irreversibly, but congelation *can* be brought about in a manner that is both isothermal and reversible. Consider the following three-step process: (1) The water is vaporized isothermally and reversibly at its equilibrium vapor pressure (P_w) at −10°C; (2) The vapor is expanded reversibly and isothermally until its pressure drops to equality with the vapor pressure over ice at −10°C; and (3) The expanded vapor is deposited on a solid ice surface by a reversible isothermal condensation at the vapor pressure (P_i) of ice at −10°C.

(a) Symbolizing by $\Delta\overline{H}_{\text{fus}}$ the heat of fusion of ice at −10°C, and bearing in mind the relation of $\Delta\overline{H}_{\text{fus}}$ to the heats of sublimation and vaporization, show that for the above tripartite process at 263°K,

$$\Delta\overline{S} = \frac{-\Delta\overline{H}_{\text{fus}}}{263} + R\ln\frac{P_w}{P_i}.$$

(b) When ice and water stand in equilibrium with each other at 273°K, their vapor pressures must be equal. By making appropriate use of the Clausius-Clapeyron equation, you can then easily demonstrate that

$$\ln\frac{P_w}{P_i} = -\frac{\Delta\overline{H}_{\text{fus}}}{R}\left[\frac{1}{273} - \frac{1}{263}\right].$$

(c) Combining (b) and (a), one obtains an expression in terms of $\Delta\overline{H}_{\text{fus}}$. Let it now be given that at 0°C the heat of fusion of ice is 1440 cal/mole, and that for ice and for water C_P is 9 and 18 cal/mole-°K respectively. From these data calculate $\Delta\overline{H}_{\text{fus}}$ at −10°C, and proceed thence to calculate $\Delta\overline{S}$ for the solidification of water at −10°C.

(d) In the actual spontaneous process the entropy change in the surroundings is clearly $-\Delta\overline{H}_{\text{fus}}/263$. Formulate a general expression for the *total* entropy change of system *and* surroundings when supercooled water freezes at *any* temperature (T^*) to which corresponds some particular value $\Delta\overline{H}^*_{\text{fus}}$.

48. (a) Consider two pure liquids (A and B) that mix to form a solution in which both conform to the standard of ideality set by equation (49). Show that the change of free energy which accompanies the mixing of a moles of pure A with b moles of pure B is given by the equation:

$$\Delta G_{\text{mix}} = aRT\ln X_A + bRT\ln X_B.$$

(b) For n moles of a polycomponent ideal solution

$$\Delta G_{\text{mix}} = nRT \sum X_i \ln X_i,$$

where X_i is the mole fraction of the ith component in the mixture, and the summation is to extend over all components. Derive this equation, and indicate how it implies that the mixing of the components is a spontaneous process.

(c) Applying, to this expression for ΔG_{mix}, the Gibbs-Helmholtz equation derived on p. 105, show that $\Delta H_{\text{mix}} = 0$.

(d) Applying, to the expression for ΔG_{mix}, the relation expressed in equation (44), show that when the mixing process is conducted isothermally $\Delta V_{\text{mix}} = 0$.

(e) Find an expression for ΔS_{mix}. Can you now say any more about how it is that the mixing process is a spontaneous one?

49. (a) The variation with temperature of the equilibrium vapor pressure of a liquid can be expressed by the equation:

$$P = \alpha e^{-\Delta H \text{vap}/RT},$$

where α is a constant characteristic of the liquid concerned. Derive this equation.

(b) Everdell notes that for a great many liquids the value of α closely approximates 2.7×10^4 atm. To what extent does Trouton's rule offer an explanation for this finding?

50. A very dilute solution of an involatile solute boils at a temperature (T) which is approximately related to the boiling temperature (T_B) of the pure solvent, by the equation

$$X = \left[\frac{T_B}{T}\right]^{10.5},$$

where X is the mole fraction of the solvent in the solution. Derive this relation by starting from equation (50), assuming the applicability of Trouton's rule, and noting that for β very close to 1 an excellent approximation is $\ln \beta = \beta - 1$.

51. (a) Other things being equal, the boiling-point-elevation constant K_B considerably exceeds the freezing-point-depression constant K_F of the same solvent. Rationalize this relationship on the basis of what you know about the magnitudes of entropies of fusion and entropies of vaporization.

(b) Justify the following statement: At any given temperature, in any solvent with which it forms an ideal solution, a given solid will have a greater solubility the lower its melting point and the smaller its heat of fusion.

52. (a) Pure p-dibromobenzene melts at 360°K, with 4.84 kcal/mole as its heat of fusion. Calculate the ideal solubility of this compound at 298°K.

(b) At 298°K a saturated solution of p-dibromobenzene (molecular weight, 236) in benzene (molecular weight, 78) contains 46% by weight of p-dibromobenzene. How good is the agreement between this empirical value and the theoretical value calculated in (a)?

53. (a) A 1% solution (amalgam) of zinc in mercury boils at 358.3°C. The heat of vaporization of mercury is 70.6 cal/gm and the boiling point of pure mercury is 356.58°C. Calculate the boiling-point-elevation constant for mercury

and, taking the atomic weight of zinc as 65.4, determine the condition of zinc dissolved in mercury (i.e., does it exist as atoms, groups of atoms, etc.?).

(b) From the slope of the line that most nearly approaches the theoretical line in Fig. 35, calculate the heat of fusion of naphthalene. Compare your result with that given in Table 6.

(c) The human red blood cell shrinks when placed in aqueous salt solution more concentrated than 0.9% NaCl and swells in a less concentrated solution. Taking 58.5 as the formula weight of NaCl (assumed 100% ionized) calculate the osmotic pressure of the solution inside the cell, at the normal body temperature of 37°C.

54. (a) Denbigh suggests that we think of the ideal solubility of gases in Raoult's-law terms. If at the temperature in question the vapor pressure of the pure condensed gas is P^0, and if P ($<P^0$) is the pressure of the gas during the measurement of its solubility, then the mole fraction (X_u) of the dissolved gas should be: $X_u = P/P^0$. Justify this formulation.

(b) From the known normal boiling point (T_B) of the gas concerned, the Clausius-Clapeyron equation will permit us to calculate P^0 at any other temperature. Even when this is a purely nominal value for P^0 at a temperature far above the critical point of the gas concerned, Denbigh suggests that the above equation yields reasonable values for the ideal solubility of the gas. If the solubility measurement is made under a pressure of 1 atmosphere, show that this entire line of analysis eventuates in the expression for X_u we have given on p. 121.

55. (a) The semipermeable membrane shown in the figure below is imagined to act in somewhat the manner of a sintered-glass disk toward mercury—retaining the liquid while passing the vapor. After the equilibrium pressure (P_G) has been established in the gas phase above the membrane, we infinitesimally increase the pressure (P_L) exerted on the liquid by the piston, while holding the temperature constant. For the change (dP_G) so produced in the vapor pressure, derive the following equation (and define $\overline{V}_L$ and $\overline{V}_G$).

$$\overline{V}_L \, dP_L = \overline{V}_G \, dP_G.$$

(b) Since $\overline{V}_L$ and $\overline{V}_G$ are always finite and positive, an increase in the pressure imposed on a liquid should always increase its vapor pressure—but moderate

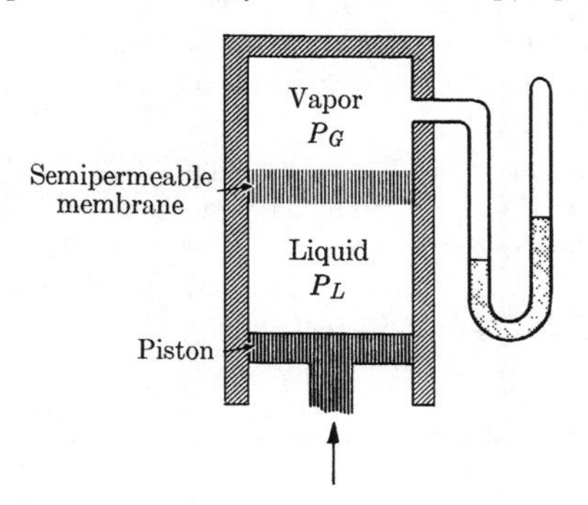

PROBLEM 55

increases in the imposed pressure do not yield directly determinable changes of vapor pressure. Why?

(c) We can regard osmotic pressure (π) as the *extra* pressure that must be imposed on a solution in order to bring its vapor pressure (P_G) up to that (P_G^0) of the pure solvent at the same temperature. Assuming (i) incompressibility of the solution, (ii) equality of the molar volume of the solvent in solution to that ($\overline{V}_L$) of the pure solvent, and (iii) ideality of the solvent vapor; integrate the equation in (a) to obtain

$$\pi \overline{V}_L = RT \ln \frac{P_G^0}{P_G}.$$

(d) Comment on the relation of the last expression to equation (54) on p. 123.

56. (a) Starting from equation (56) as the definition of an ideal gas, show that the equation of state for such a gas is $P\overline{V} = RT$.

(b) Show that in a mixture of ideal gases, where the total pressure is P, the mole fraction (X_i) of the ith component is expressible in terms of its partial pressure (P_i), as $X_i = P_i/P$.

(c) Show that in a mixture of ideal gases the molar concentration (c_i) of the ith component is expressible in terms of its partial pressure, as $c_i = P_i/RT$.

57. For our type-reaction of *ideal gases*, $aA + bB = zZ + yY$, the equilibrium constant $K_P[=(P_Z)^z(P_Y)^y/(P_A)^a(P_B)^b]$ is a useful function of temperature *only*.

(a) An occasionally useful equilibrium constant, expressed in terms of mole fractions, is defined by writing $K_X = (X_Z)^z(X_Y)^y/(X_A)^a(X_B)^b$. Keeping in view the relation stated in problem 56(b), show that

$$K_X = K_P P^{-\Delta n},$$

where Δn symbolizes the change in the number of moles of gas consequent to the reaction as written, and P is the total pressure.

(b) Unlike K_P, K_X is clearly a function of the total pressure—save in the special case that $\Delta n = 0$. Keeping in mind that K_P *is* independent of pressure, show that the equation in (a) entails that, for our reaction of ideal gases, the pressure-dependence of K_X be expressible as

$$\left[\frac{d \ln K_X}{dP}\right]_T = -\frac{\Delta V}{RT},$$

where ΔV is the change of volume consequent to the reaction as written.

(c) Comment on the relation of the last expression to (i) Le Chatelier's principle, and (ii) equation (44).

58. For our type-reaction of *ideal gases*, $aA + bB = zZ + yY$, in terms of partial pressures we write as the equilibrium constant $K_P = (P_Z)^z(P_Y)^y/(P_A)^a(P_B)^b$.

(a) In terms of molar concentrations (c), an occasionally useful equilibrium constant is defined by writing $K_c = (c_Z)^z(c_Y)^y/(c_A)^a(c_B)^b$. Keeping in view the relation stated in problem 56(c), show that

$$K_c = K_P(RT)^{-\Delta n},$$

where Δn is the change in the number of moles of gas accompanying the reaction as written.

(b) Knowing that equation (64) applies to K_P, for a reaction of ideal gases derive from the equation in (a) the following variant of van't Hoff's law:

$$\frac{d \ln K_c}{dT} = \frac{\Delta E^0}{RT^2},$$

where ΔE^0 is the standard energy change consequent to the reaction.

(c) Mole fractions don't change with volume, while molar concentrations do so change. Yet K_X is dependent on pressure while K_c is not. Explain!

59. For reactions in solution, the fundamentally important equilibrium constant is that (K_X) in terms of mole fractions. But for very dilute solutions we may find it convenient to use the equilibrium constants K_c and K_m, expressed respectively in molar and molal concentrations.

(a) Show that in very dilute solutions the molar concentration (c_i) and molality (m_i) of the ith solute are related to its mole fraction (X_i) by the equations

$$c_i \cong 1000\, X_i/\overline{V}_s \quad \text{and} \quad m_i \cong 1000 X_i/M_s,$$

where $\overline{V}_s$ and M_s respectively symbolize the molar volume (in ml) and the gram-molecular weight of the solvent.

(b) For the type-reaction $aA + bB = zZ + yY$, proceeding in a highly dilute ideal solution, we write:

$$K_c = (c_Z)^z(c_Y)^y/(c_A)^a(c_B)^b \quad \text{and} \quad K_m = (m_Z)^z(m_Y)^y/(m_A)^a(m_B)^b.$$

Show that K_c and K_m are related to K_X by the equations:

$$K_c = K_X(\overline{V}_s/1000)^{-\Delta n} \quad \text{and} \quad K_m = K_X(M_s/1000)^{-\Delta n},$$

where Δn symbolizes the change in number of moles consequent to the reaction as written.

(c) How do the last equations imply that, like K_X, K_c and K_m will be constants for a given reaction at a given temperature in a given solvent? Actually, unlike K_X, K_c and K_m are acceptably constant only in the very highly dilute solutions for which the approximation used in (a) is permissible. Considering the nature of that approximation, explain why the high-dilution restriction is notably less severe in the special case of aqueous solutions.

60. In the text we define a standard free energy ($\overline{G}_\ell^0$) for a dissolved substance in the standard state for which the mole fraction $X = 1$. Sometimes we find it convenient to use a standard free energy $\overline{G}_\ell^c$ defined for a standard state in which the molar concentration $c = 1$, or a standard free energy $\overline{G}_\ell^m$ defined for a standard state in which the molal concentration $m = 1$.

(a) Setting out from equation (49), and noting the relations stated in problem 59(a), show that:

$$\overline{G}_\ell = \overline{G}_\ell^c + RT \ln c, \quad \text{where} \quad \overline{G}_\ell^c = \overline{G}_\ell^0 + RT \ln (\overline{V}_s/1000),$$

and

$$\overline{G}_\ell = \overline{G}_\ell^m + RT \ln m, \quad \text{where} \quad \overline{G}_\ell^m = \overline{G}_\ell^0 + RT \ln (M_s/1000),$$

and $\overline{G}_\ell$ represents the molar free energy of the species at the indicated molarity or molality.

(b) When determined by extrapolation from highly dilute solutions, $\overline{G}_{l}^{c}$ and $\overline{G}_{l}^{m}$ refer to strictly *hypothetical* standard states in which, at unit concentration, the substance would still display the qualities it shows at high dilution. However hypothetical, these are nonetheless genuine *standard* states: for a given solute in a given solvent, $\overline{G}_{l}^{c}$ and $\overline{G}_{l}^{m}$ are constants dependent only on temperature. From them we can then derive equilibrium constants K_c and K_m—just as we formerly derived K_X from consideration of $\overline{G}_{l}^{0}$. Perform the derivations, and show that K_c and K_m so derived are related to K_X by the equations stated in problem 59(b).

(c) While in an ideal solution K_X remains the same at all concentrations, K_c and K_m remain constant only in the range of very low concentrations. Where has this restriction entered into your derivation of K_c and K_m?

61. (a) An important operation in petroleum refining is the gas-phase isomerization of *n*-butane, $CH_3CH_2CH_2CH_3$, to *i*-butane, $CH_3CH(CH_3)_2$. Relevant thermodynamic data are:

	n-butane (g)	*i*-butane (g)
ΔH^0_{f298}, kcal/mole	−30.15	−32.15
S^0_{298}, cal/mole-°K	+74.12	+70.42

Calculate K_P and K_c for the isomerization reaction at 298°K and 1 atm pressure.

(b) Would you expect the yield of *i*-butane to be improved by increasing the pressure? by increasing the temperature?

(c) As a rule of thumb for simple gas-phase reactions at room temperature, Roberts and Caserio suggest that "K usually is greater than unity for most reactions with ΔH^0 more negative than −15 kcal and is usually less than unity for ΔH^0 more positive than +15 kcal." What does this somewhat ambiguous statement suggest about the "usual" order of magnitude of ΔS^0 in "most reactions"?

62. (a) Consider the following data for the standard states of the indicated materials at 298°K:

	NH_4NO_3 (s)	H_2O (g)
ΔH^0_f (kcal/mole)	−87.3	−57.8
ΔG^0_f (kcal/mole)	−45.1	−54.6

Once one envisions the possibility of the reaction

$$NH_4NO_3\ (s) = N_2\ (g) + \tfrac{1}{2}O_2\ (g) + 2H_2O\ (g),$$

the above data are, to the initiate, another way of spelling DANGER. Explain!

(b) At 298°K the free energy of formation of NO (g) is $\Delta G^0_{f298} = +20.7$ kcal/mole. How is it possible that NO may exist at 298°K? How is it possible that NO can be formed at 298°K and 1 atm pressure?

63. (a) Derive the Nernst equation by the method indicated in the footnote on p. 132.

(b) The Daniell cell involves the reaction

$$Zn\ (s) + Cu^{++}\ (aq) = Zn^{++}\ (aq) + Cu\ (s).$$

At 273°K, $\mathfrak{E}^0 = 1.093$ volts for this cell, which has a temperature coefficient of

-4.53×10^{-4} volt/°K. For the above reaction the value of ΔH^0 determined by direct calorimetry is —55.2 kcal; calculate a value of ΔH^0 from the electrochemical data.

64. (a) Derive the expression: $\dfrac{\Delta G}{\Delta H} = 1 - T\dfrac{\Delta S}{\Delta H}\cdot$

(b) When a spontaneous exothermic reaction is brought about by direct mixing of suitable components, ΔH expresses the magnitude of the heat output to the surroundings. When the same reaction is brought about by reversible operation of a galvanic cell containing the same components, ΔG expresses the magnitude of the work output to the surroundings. Taking due account of the signs of ΔG and ΔH, use the equation in (a) to determine under what conditions $\Delta G/\Delta H > 1$.

(c) When, of two initially identical systems, one proceeds through the irreversible change, and the other through the reversible change noted under (b), both may appear at the end in exactly the same state. Yet from the reversible operation of the galvanic cell we may recover a work output ($-\Delta G$) that substantially exceeds the heat output ($-\Delta H$) from the irreversible reaction. How is this at all reconcilable with the first principle of thermodynamics?

(d) For CO, at 298°K, $\Delta H_f^0 = -26.42$ kcal/mole, and $\Delta G_f^0 = -32.78$ kcal/mole. For the reaction C (s) + O_2 (g) = 2CO (g), determine the ratio $\Delta G^0/\Delta H^0$ —corresponding to the ratio of (i) the work output from a reversibly operated fuel cell to (ii) the heat output obtained by direct (irreversible) combination of the reactants at 298°K.

65. (a) From the experimental results plotted in Fig. 38, calculate (with due attention to units) the values of ΔG^0, ΔH^0, and ΔS^0 for the following reaction at 150°C:

$$(CH_3COOH)_2 \text{ (g)} = 2CH_3COOH \text{ (g)}.$$

(b) In a reaction of the type $A \rightarrow 2X$, the standard free-energy change and the equilibrium constant are correlated by the equation $RT \ln K = -\Delta G^0$. Show how the same equation continues to correlate the different values of ΔG^0 and K that correspond to the following two different expressions of the same reaction:

$$\tfrac{1}{2}A \rightarrow X \qquad \text{and} \qquad 2X \rightarrow A.$$

66. With reference to the reaction

$$CaCO_3 \text{ (s)} = CaO \text{ (s)} + CO_2 \text{ (g)},$$

consider the following data for the standard states of the indicated materials at 298°K:

	$CaCO_3$ (s)	CaO (s)	CO_2 (g)
ΔH_f^0 (kcal/mole)	—288.5	—151.9	—94.05
S_{298}^0 (cal/mole-°K)	+22.2	+9.5	+51.06

(a) Determine the equilibrium constant for the reaction at 298°K, and state its units.

(b) On the (good) assumption that ΔH^0 for the reaction is substantially constant over the temperature range concerned, calculate the temperature at which the equilibrium pressure of CO_2 becomes equal to 1 atm.

(c) Reviewing the way in which we arrived at the concept of equilibrium constant, explain why one may expect that the equilibrium pressure of CO_2 at a given temperature will be independent of the proportion of CaO to $CaCO_3$.

67. (a) From equation (11), and from the corresponding equation on p. 89, derive the following two relations:

$$\left[\frac{d\,\Delta H^0}{dT}\right]_P = \Delta C_P \qquad \text{and} \qquad \left[\frac{d\,\Delta S^0}{dT}\right]_P = \frac{\Delta C_P}{T}.$$

(b) How do these equations justify the indication (on p. 134) that, if ΔH^0 is effectively constant over some range of temperature, ΔS^0 for the given change will be even better approximated as a constant over the same range.

(c) On p. 98 it is argued that, if ΔH^0 and ΔS^0 are alike in sign, acceptable values for ΔG^0 (and K) can usually be obtained by treating ΔH^0 and ΔS^0 as constant even over a range of temperature in which ΔC_P is substantial in magnitude. Considering the relation of ΔH^0 and ΔS^0 to ΔG^0 (and thence to K) justify this contention.

68. (a) Over a short enough span of temperature, ΔH^0 is certainly well enough approximated as a constant to justify use of the integral form of van't Hoff's law expressed in equation (63). A simple though inelegant derivation of the differential form expressed in equation (64) then follows from application of equation (63) to the special case in which the two temperatures differ only infinitesimally. Setting T_1 and T_2 equal to T and $T + dT$ respectively, perform the derivation of equation (64) from equation (63).

(b) Treating K, T, ΔH^0, and ΔS^0 all as variables, differentiate equation (62) with respect to temperature. Noting then the relations expressed in problem 67(a), complete the derivation of equation (64) from equation (62).

69. The Deacon process, once used for the manufacture of Cl_2 from by-product HCl, depends on a reaction that does not proceed at a detectable rate at low temperatures and that has an unfavorable equilibrium at high temperatures. The reaction is this: $4HCl\,(g) + O_2\,(g) = 2H_2O\,(g) + 2Cl_2\,(g)$. The optimum operating temperature is said to be 470°C. Calculate the equilibrium constant for the reaction at that temperature, given the following data for the standard states of the indicated materials at 298°K:

	HCl (g)	O_2 (g)	H_2O (g)	Cl_2 (g)
ΔH_f^0 (kcal/mole)	−22.1	—	−57.8	—
S_{298}^0 (cal/mole-°K)	+44.6	+49.0	+45.1	+53.3

Appendix IV

*Thermochemical Data at 298.15°K**

Values are given for ΔH_f^0, the standard enthalpy of formation; ΔG_f^0, the standard free energy of formation; S^0, the standard (Nernst theorem) entropy; and C_P, the heat capacity at constant pressure—in each case for one mole of material in the indicated state at 1 atm pressure and 25°C. The symbol (aq) refers to a hypothetical ideal aqueous solution in which, at unit molality, the indicated ion has the molal enthalpy and heat capacity one finds for it by extrapolating to infinite dilution. The values for the corresponding ideal salt solutions of unit molality are obtained simply by making the appropriate sums of the values for the component ions. The listed values for ions are based on the convention that takes the value of all the listed properties to be 0.0 for H^+ (aq).

Substance	ΔH_f^0 kcal/mole	ΔG_f^0 kcal/mole	S^0 cal/mole-°K	C_P cal/mole-°K
Ag (s)	0.00	0.00	10.206	6.092
AgBr (s)	−23.78	−22.930	25.60	12.52
AgCl (s)	−30.362	−26.224	22.97	12.14
AgI (s)	−14.91	−15.85	27.3	13.01
Al (s)	0.00	0.00	6.77	5.82
Al_2O_3 (s, corundum)	−400.5	−378.2	12.17	18.89
Br (g)	26.741	19.701	41.805	4.968
Br^- (aq)	−29.05	−24.85	19.7	−33.9
BrCl (g)	3.50	−0.23	57.36	8.36
Br_2 (g)	7.387	0.751	58.641	8.61
$Br_2(\ell)$	0.00	0.00	36.384	18.090
C (g)	171.291	160.442	37.7597	4.9805
C (s, diamond)	0.4533	0.6930	0.568	1.4615
C (s, graphite)	0.00	0.00	1.372	2.038
CCl_4 (g)	−24.6	−14.49	74.03	19.91

* From *Selected Values of Chemical Thermodynamic Properties*, issued by the National Bureau of Standards. The earlier complete tabulation appearing in 1952 as NBS Circular 500, ed. F. D. Rossini *et al.*, has been used only to fill lacunae in the uncompleted tabulation appearing in 1968 as NBS Technical Note 270–3, ed. D. D. Wagman *et al.*

Substance	ΔH_f^0 kcal/mole	ΔG_f^0 kcal/mole	S^0 cal/mole-°K	C_P cal/mole-°K
CH_4 (g)	−17.88	−12.13	44.492	8.439
CO (g)	−26.416	−32.780	47.219	6.959
CO_2 (g)	−94.051	−94.254	51.06	8.87
CS_2 (g)	28.05	16.05	56.82	10.85
C_2H_2 (g)	54.19	50.00	48.00	10.50
C_2H_4 (g)	12.49	16.28	52.45	10.41
C_2H_6 (g)	−20.24	−7.86	54.85	12.58
C_2N_2 (g)	73.84	71.07	57.79	13.58
Ca (s)	0.00	0.00	9.95	6.28
Ca^{++} (aq)	−129.77	−132.18	−13.2	
$CaCO_3$ (calcite)	−288.45	−269.78	22.2	19.57
$CaCO_3$ (aragonite)	−288.49	−269.53	21.2	19.42
CaC_2 (s)	−15.0	−16.2	16.8	14.90
$CaCl_2$ (s)	−190.0	−179.3	27.2	17.36
CaO (s)	−151.9	−144.4	9.5	10.23
$Ca(OH)_2$ (s)	−235.80	−214.33	18.2	20.2
Cl (g)	29.082	25.262	39.457	5.220
Cl^- (aq)	−39.952	−31.372	13.5	−32.6
Cl_2 (g)	0.00	0.00	53.288	8.104
Cu (s)	0.00	0.00	7.96	5.848
CuCl (s)	−32.2	−28.4	21.9	
$CuCl_2$ (s)	−49.2			
CuO (s)	−37.1	−30.4	10.4	10.6
Cu_2O (s)	−39.84	−34.98	24.1	16.7
Fe (s)	0.00	0.00	6.49	6.03
Fe_2O_3 (s)	−196.5	−177.1	21.5	25.0
Fe_3O_4 (s)	−267.0	−242.4	35.0	
H (g)	52.095	48.581	27.391	4.9679
H^+ (aq)	0.00	0.00	0.00	0.00
HBr (g)	−8.70	−12.77	47.463	6.965
HCl (g)	−22.062	−22.777	44.646	6.96
HI (g)	6.33	0.41	49.351	6.969
H_2 (g)	0.00	0.00	31.208	6.889
H_2O (g)	−57.796	−54.634	45.104	8.025
H_2O (ℓ)	−68.315	−56.687	16.71	17.995
H_2O_2 (ℓ)	−44.88	−28.78	26.2	21.3
H_2S (g)	−4.93	−8.02	49.16	8.18
Hg (g)	14.54	7.59	41.80	4.968
Hg (ℓ)	0.00	0.00	18.5	6.65
$HgCl_2$ (s)	−55.0			18.3

Substance	ΔH_f^0 kcal/mole	ΔG_f^0 kcal/mole	S^0 cal/mole-°K	C_P cal/mole-°K
HgO (s, red)	−21.68	−13.990	17.2	10.93
HgO (s, yellow)	−21.56	−13.959	17.5	
Hg_2Cl_2 (s)	−63.32	−50.350	46.8	24.3
I (g)	25.535	16.798	43.184	4.968
I^- (aq)	−13.19	−12.33	26.6	−34.0
I_2 (g)	14.923	4.627	62.28	8.82
I_2 (s)	0.00	0.00	27.757	13.011
I_3^- (aq)	−12.3	−12.3	57.2	
K (s)	0.00	0.00	15.2	6.97
K^+ (aq)	−60.04	−67.466	24.5	
KBr (s)	−93.73	−90.63	23.05	12.82
KCl (s)	−104.175	−97.592	19.76	12.31
KI (s)	−78.31	−77.03	24.94	13.16
Mg (s)	0.00	0.00	7.77	5.71
Mg^{++} (aq)	−110.41	−108.99	−28.2	
$MgCl_2$ (s)	−153.40	−141.57	21.4	17.04
MgO (s)	−143.84	−136.13	6.4	8.94
$Mg(OH)_2$ (s)	−221.00	−199.27	15.09	18.41
NH_3 (g)	−11.02	−3.94	45.97	8.38
NH_4^+ (aq)	−31.67	−18.97	27.1	19.1
NO (g)	21.57	20.69	50.347	7.133
NO_2 (g)	7.93	12.26	57.35	8.89
NO_2^- (aq)	−25.0	−8.9	33.5	−23.3
NO_3^- (aq)	−49.56	−26.61	35.0	−20.7
N_2 (g)	0.00	0.00	45.77	6.961
N_2O (g)	19.61	24.90	52.52	9.19
N_2O_4 (g)	2.19	23.38	72.70	18.47
Na (s)	0.00	0.00	12.2	6.79
Na^+ (aq)	−57.279	−62.589	14.4	
NaBr (s)	−86.030			12.5
NaCl (s)	−98.232	−91.785	17.30	11.88
$NaHCO_3$ (s)	−226.5	−203.6	24.4	20.94
NaOH (s)	−101.99		32.5	19.2
Na_2CO_3 (s)	−270.3	−250.4		26.41
OH^- (aq)	−54.970	−37.594	−2.57	−35.5
O_2 (g)	0.00	0.00	49.003	7.016
O_3 (g)	34.1	39.0	57.08	9.37
Pb (s)	0.00	0.00	15.49	6.32
PbO (s, yellow)	−51.94	−44.91	16.42	10.94
PbO_2 (s)	−66.3	−51.95	16.4	15.45
$PbSO_4$ (s)	−219.87	−194.36	35.51	24.667
Pb_3O_4 (s)	−171.7	−143.7	50.5	35.1

Substance	ΔH_f^0 kcal/mole	ΔG_f^0 kcal/mole	S^0 cal/mole-°K	C_P cal/mole-°K
S (s, rhombic)	0.00	0.00	7.60	5.41
S (s, monoclinic)	0.071	0.023	7.78	5.65
SO_2 (g)	−70.944	−71.748	59.30	9.53
SO_3 (g)	−94.58	−88.69	61.34	12.11
$SO_4^=$ (aq)	−217.32	−177.97	4.8	−70.
S_8 (g)	24.45	11.87	102.98	37.39

Molar Heat Capacities from 298° to 2500°K*

When C_P is expressed as the empirical power series $C_P = a + bT + cT^{-2}$, some values of the constants are as tabulated here.

Substance	a	$b \times 10^3$	$c \times 10^{-5}$
C (s, graphite)	4.03	1.14	−2.04
CH_4 (g), to 1500°K	5.65	11.44	−0.46
CO (g)	6.79	0.98	−0.11
CO_2 (g)	10.57	2.10	−2.06
H_2 (g)	6.52	0.78	0.12
H_2O (g)	7.30	2.46	
N_2 (g)	6.83	0.90	−0.12
NH_3 (g), to 1800°K	7.11	6.00	−0.37
NO (g)	7.03	0.92	−0.14
O_2 (g)	7.16	1.00	−0.40

* From K. K. Kelley, *U. S. Bur. Mines Bull.* 584 (1960).

Bond Enthalpies at 298°K*
kcal/mole

H—H	104	C—C	83	N—N	39
H—C	99	C=C	146	N≡N	226
H—N	93	C≡C	200	O—O	34
H—O	111	C—N	72	O=O	119
H—F	135	C≡N	213	O—Cl	52
H—Cl	103	C—O	85	F—F	37
H—Br	87	C=O	177	Cl—Cl	58
H—I	71	C—F	116	Cl—Br	52
		C—Cl	80	Br—Br	46
		C—Br	67	I—I	36

* To within 1 kcal/mole, all figures are those given by T. L. Cottrell in *The Strengths of Chemical Bonds*, 2nd ed. (Butterworths, London, 1958).

Index

A CATALOG OF SELECTED

DOVER BOOKS

IN SCIENCE AND MATHEMATICS

Astronomy

BURNHAM'S CELESTIAL HANDBOOK, Robert Burnham, Jr. Thorough guide to the stars beyond our solar system. Exhaustive treatment. Alphabetical by constellation: Andromeda to Cetus in Vol. 1; Chamaeleon to Orion in Vol. 2; and Pavo to Vulpecula in Vol. 3. Hundreds of illustrations. Index in Vol. 3. 2,000pp. 6⅛ x 9¼.
Vol. I: 23567-X
Vol. II: 23568-8
Vol. III: 23673-0

EXPLORING THE MOON THROUGH BINOCULARS AND SMALL TELESCOPES, Ernest H. Cherrington, Jr. Informative, profusely illustrated guide to locating and identifying craters, rills, seas, mountains, other lunar features. Newly revised and updated with special section of new photos. Over 100 photos and diagrams. 240pp. 8¼ x 11. 24491-1

THE EXTRATERRESTRIAL LIFE DEBATE, 1750–1900, Michael J. Crowe. First detailed, scholarly study in English of the many ideas that developed from 1750 to 1900 regarding the existence of intelligent extraterrestrial life. Examines ideas of Kant, Herschel, Voltaire, Percival Lowell, many other scientists and thinkers. 16 illustrations. 704pp. 5⅜ x 8½. 40675-X

THEORIES OF THE WORLD FROM ANTIQUITY TO THE COPERNICAN REVOLUTION, Michael J. Crowe. Newly revised edition of an accessible, enlightening book recreates the change from an earth-centered to a sun-centered conception of the solar system. 242pp. 5⅜ x 8½. 41444-2

A HISTORY OF ASTRONOMY, A. Pannekoek. Well-balanced, carefully reasoned study covers such topics as Ptolemaic theory, work of Copernicus, Kepler, Newton, Eddington's work on stars, much more. Illustrated. References. 521pp. 5⅜ x 8½.
65994-1

A COMPLETE MANUAL OF AMATEUR ASTRONOMY: Tools and Techniques for Astronomical Observations, P. Clay Sherrod with Thomas L. Koed. Concise, highly readable book discusses: selecting, setting up and maintaining a telescope; amateur studies of the sun; lunar topography and occultations; observations of Mars, Jupiter, Saturn, the minor planets and the stars; an introduction to photoelectric photometry; more. 1981 ed. 124 figures. 26 halftones. 37 tables. 335pp. 6½ x 9¼.
42820-6

AMATEUR ASTRONOMER'S HANDBOOK, J. B. Sidgwick. Timeless, comprehensive coverage of telescopes, mirrors, lenses, mountings, telescope drives, micrometers, spectroscopes, more. 189 illustrations. 576pp. 5⅝ x 8¼. (Available in U.S. only.)
24034-7

STARS AND RELATIVITY, Ya. B. Zel'dovich and I. D. Novikov. Vol. 1 of *Relativistic Astrophysics* by famed Russian scientists. General relativity, properties of matter under astrophysical conditions, stars, and stellar systems. Deep physical insights, clear presentation. 1971 edition. References. 544pp. 5⅝ x 8¼. 69424-0

Chemistry

THE SCEPTICAL CHYMIST: The Classic 1661 Text, Robert Boyle. Boyle defines the term "element," asserting that all natural phenomena can be explained by the motion and organization of primary particles. 1911 ed. viii+232pp. 5⅜ x 8½. 42825-7

RADIOACTIVE SUBSTANCES, Marie Curie. Here is the celebrated scientist's doctoral thesis, the prelude to her receipt of the 1903 Nobel Prize. Curie discusses establishing atomic character of radioactivity found in compounds of uranium and thorium; extraction from pitchblende of polonium and radium; isolation of pure radium chloride; determination of atomic weight of radium; plus electric, photographic, luminous, heat, color effects of radioactivity. ii+94pp. 5⅜ x 8½. 42550-9

CHEMICAL MAGIC, Leonard A. Ford. Second Edition, Revised by E. Winston Grundmeier. Over 100 unusual stunts demonstrating cold fire, dust explosions, much more. Text explains scientific principles and stresses safety precautions. 128pp. 5⅜ x 8½. 67628-5

THE DEVELOPMENT OF MODERN CHEMISTRY, Aaron J. Ihde. Authoritative history of chemistry from ancient Greek theory to 20th-century innovation. Covers major chemists and their discoveries. 209 illustrations. 14 tables. Bibliographies. Indices. Appendices. 851pp. 5⅜ x 8½. 64235-6

CATALYSIS IN CHEMISTRY AND ENZYMOLOGY, William P. Jencks. Exceptionally clear coverage of mechanisms for catalysis, forces in aqueous solution, carbonyl- and acyl-group reactions, practical kinetics, more. 864pp. 5⅜ x 8½. 65460-5

ELEMENTS OF CHEMISTRY, Antoine Lavoisier. Monumental classic by founder of modern chemistry in remarkable reprint of rare 1790 Kerr translation. A must for every student of chemistry or the history of science. 539pp. 5⅜ x 8½. 64624-6

THE HISTORICAL BACKGROUND OF CHEMISTRY, Henry M. Leicester. Evolution of ideas, not individual biography. Concentrates on formulation of a coherent set of chemical laws. 260pp. 5⅜ x 8½. 61053-5

A SHORT HISTORY OF CHEMISTRY, J. R. Partington. Classic exposition explores origins of chemistry, alchemy, early medical chemistry, nature of atmosphere, theory of valency, laws and structure of atomic theory, much more. 428pp. 5⅜ x 8½. (Available in U.S. only.) 65977-1

GENERAL CHEMISTRY, Linus Pauling. Revised 3rd edition of classic first-year text by Nobel laureate. Atomic and molecular structure, quantum mechanics, statistical mechanics, thermodynamics correlated with descriptive chemistry. Problems. 992pp. 5⅜ x 8½. 65622-5

FROM ALCHEMY TO CHEMISTRY, John Read. Broad, humanistic treatment focuses on great figures of chemistry and ideas that revolutionized the science. 50 illustrations. 240pp. 5⅜ x 8½. 28690-8

Engineering

DE RE METALLICA, Georgius Agricola. The famous Hoover translation of greatest treatise on technological chemistry, engineering, geology, mining of early modern times (1556). All 289 original woodcuts. 638pp. 6¾ x 11. 60006-8

FUNDAMENTALS OF ASTRODYNAMICS, Roger Bate et al. Modern approach developed by U.S. Air Force Academy. Designed as a first course. Problems, exercises. Numerous illustrations. 455pp. 5⅜ x 8½. 60061-0

DYNAMICS OF FLUIDS IN POROUS MEDIA, Jacob Bear. For advanced students of ground water hydrology, soil mechanics and physics, drainage and irrigation engineering, and more. 335 illustrations. Exercises, with answers. 784pp. 6⅛ x 9¼. 65675-6

THEORY OF VISCOELASTICITY (Second Edition), Richard M. Christensen. Complete, consistent description of the linear theory of the viscoelastic behavior of materials. Problem-solving techniques discussed. 1982 edition. 29 figures. xiv+364pp. 6⅛ x 9¼. 42880-X

MECHANICS, J. P. Den Hartog. A classic introductory text or refresher. Hundreds of applications and design problems illuminate fundamentals of trusses, loaded beams and cables, etc. 334 answered problems. 462pp. 5⅜ x 8½. 60754-2

MECHANICAL VIBRATIONS, J. P. Den Hartog. Classic textbook offers lucid explanations and illustrative models, applying theories of vibrations to a variety of practical industrial engineering problems. Numerous figures. 233 problems, solutions. Appendix. Index. Preface. 436pp. 5⅜ x 8½. 64785-4

STRENGTH OF MATERIALS, J. P. Den Hartog. Full, clear treatment of basic material (tension, torsion, bending, etc.) plus advanced material on engineering methods, applications. 350 answered problems. 323pp. 5⅜ x 8½. 60755-0

A HISTORY OF MECHANICS, René Dugas. Monumental study of mechanical principles from antiquity to quantum mechanics. Contributions of ancient Greeks, Galileo, Leonardo, Kepler, Lagrange, many others. 671pp. 5⅜ x 8½. 65632-2

STABILITY THEORY AND ITS APPLICATIONS TO STRUCTURAL MECHANICS, Clive L. Dym. Self-contained text focuses on Koiter postbuckling analyses, with mathematical notions of stability of motion. Basing minimum energy principles for static stability upon dynamic concepts of stability of motion, it develops asymptotic buckling and postbuckling analyses from potential energy considerations, with applications to columns, plates, and arches. 1974 ed. 208pp. 5⅜ x 8½. 42541-X

METAL FATIGUE, N. E. Frost, K. J. Marsh, and L. P. Pook. Definitive, clearly written, and well-illustrated volume addresses all aspects of the subject, from the historical development of understanding metal fatigue to vital concepts of the cyclic stress that causes a crack to grow. Includes 7 appendixes. 544pp. 5⅜ x 8½. 40927-9

Mathematics

FUNCTIONAL ANALYSIS (Second Corrected Edition), George Bachman and Lawrence Narici. Excellent treatment of subject geared toward students with background in linear algebra, advanced calculus, physics, and engineering. Text covers introduction to inner-product spaces, normed, metric spaces, and topological spaces; complete orthonormal sets, the Hahn-Banach Theorem and its consequences, and many other related subjects. 1966 ed. 544pp. 6⅛ x 9¼. 40251-7

ASYMPTOTIC EXPANSIONS OF INTEGRALS, Norman Bleistein & Richard A. Handelsman. Best introduction to important field with applications in a variety of scientific disciplines. New preface. Problems. Diagrams. Tables. Bibliography. Index. 448pp. 5⅜ x 8½. 65082-0

VECTOR AND TENSOR ANALYSIS WITH APPLICATIONS, A. I. Borisenko and I. E. Tarapov. Concise introduction. Worked-out problems, solutions, exercises. 257pp. 5⅝ x 8¼. 63833-2

THE ABSOLUTE DIFFERENTIAL CALCULUS (CALCULUS OF TENSORS), Tullio Levi-Civita. Great 20th-century mathematician's classic work on material necessary for mathematical grasp of theory of relativity. 452pp. 5⅝ x 8¼. 63401-9

AN INTRODUCTION TO ORDINARY DIFFERENTIAL EQUATIONS, Earl A. Coddington. A thorough and systematic first course in elementary differential equations for undergraduates in mathematics and science, with many exercises and problems (with answers). Index. 304pp. 5⅜ x 8½. 65942-9

FOURIER SERIES AND ORTHOGONAL FUNCTIONS, Harry F. Davis. An incisive text combining theory and practical example to introduce Fourier series, orthogonal functions and applications of the Fourier method to boundary-value problems. 570 exercises. Answers and notes. 416pp. 5⅜ x 8½. 65973-9

COMPUTABILITY AND UNSOLVABILITY, Martin Davis. Classic graduate-level introduction to theory of computability, usually referred to as theory of recurrent functions. New preface and appendix. 288pp. 5⅜ x 8½. 61471-9

ASYMPTOTIC METHODS IN ANALYSIS, N. G. de Bruijn. An inexpensive, comprehensive guide to asymptotic methods–the pioneering work that teaches by explaining worked examples in detail. Index. 224pp. 5⅜ x 8½ 64221-6

APPLIED COMPLEX VARIABLES, John W. Dettman. Step-by-step coverage of fundamentals of analytic function theory–plus lucid exposition of five important applications: Potential Theory; Ordinary Differential Equations; Fourier Transforms; Laplace Transforms; Asymptotic Expansions. 66 figures. Exercises at chapter ends. 512pp. 5⅜ x 8½. 64670-X

INTRODUCTION TO LINEAR ALGEBRA AND DIFFERENTIAL EQUATIONS, John W. Dettman. Excellent text covers complex numbers, determinants, orthonormal bases, Laplace transforms, much more. Exercises with solutions. Undergraduate level. 416pp. 5⅜ x 8½. 65191-6

INTRODUCTORY REAL ANALYSIS, A.N. Kolmogorov, S. V. Fomin. Translated by Richard A. Silverman. Self-contained, evenly paced introduction to real and functional analysis. Some 350 problems. 403pp. 5⅜ x 8½. 61226-0

APPLIED ANALYSIS, Cornelius Lanczos. Classic work on analysis and design of finite processes for approximating solution of analytical problems. Algebraic equations, matrices, harmonic analysis, quadrature methods, more. 559pp. 5⅜ x 8½. 65656-X

AN INTRODUCTION TO ALGEBRAIC STRUCTURES, Joseph Landin. Superb self-contained text covers "abstract algebra": sets and numbers, theory of groups, theory of rings, much more. Numerous well-chosen examples, exercises. 247pp. 5⅜ x 8½. 65940-2

QUALITATIVE THEORY OF DIFFERENTIAL EQUATIONS, V. V. Nemytskii and V.V. Stepanov. Classic graduate-level text by two prominent Soviet mathematicians covers classical differential equations as well as topological dynamics and ergodic theory. Bibliographies. 523pp. 5⅜ x 8½. 65954-2

THEORY OF MATRICES, Sam Perlis. Outstanding text covering rank, nonsingularity and inverses in connection with the development of canonical matrices under the relation of equivalence, and without the intervention of determinants. Includes exercises. 237pp. 5⅜ x 8½. 66810-X

INTRODUCTION TO ANALYSIS, Maxwell Rosenlicht. Unusually clear, accessible coverage of set theory, real number system, metric spaces, continuous functions, Riemann integration, multiple integrals, more. Wide range of problems. Undergraduate level. Bibliography. 254pp. 5⅜ x 8½. 65038-3

MODERN NONLINEAR EQUATIONS, Thomas L. Saaty. Emphasizes practical solution of problems; covers seven types of equations. ". . . a welcome contribution to the existing literature. . . . "–*Math Reviews*. 490pp. 5⅜ x 8½. 64232-1

MATRICES AND LINEAR ALGEBRA, Hans Schneider and George Phillip Barker. Basic textbook covers theory of matrices and its applications to systems of linear equations and related topics such as determinants, eigenvalues, and differential equations. Numerous exercises. 432pp. 5⅜ x 8½. 66014-1

MATHEMATICS APPLIED TO CONTINUUM MECHANICS, Lee A. Segel. Analyzes models of fluid flow and solid deformation. For upper-level math, science, and engineering students. 608pp. 5⅜ x 8½. 65369-2

ELEMENTS OF REAL ANALYSIS, David A. Sprecher. Classic text covers fundamental concepts, real number system, point sets, functions of a real variable, Fourier series, much more. Over 500 exercises. 352pp. 5⅜ x 8½. 65385-4

SET THEORY AND LOGIC, Robert R. Stoll. Lucid introduction to unified theory of mathematical concepts. Set theory and logic seen as tools for conceptual understanding of real number system. 496pp. 5⅜ x 8¼. 63829-4

Math–Decision Theory, Statistics, Probability

ELEMENTARY DECISION THEORY, Herman Chernoff and Lincoln E. Moses. Clear introduction to statistics and statistical theory covers data processing, probability and random variables, testing hypotheses, much more. Exercises. 364pp. 5⅜ x 8½. 65218-1

STATISTICS MANUAL, Edwin L. Crow et al. Comprehensive, practical collection of classical and modern methods prepared by U.S. Naval Ordnance Test Station. Stress on use. Basics of statistics assumed. 288pp. 5⅜ x 8½. 60599-X

SOME THEORY OF SAMPLING, William Edwards Deming. Analysis of the problems, theory, and design of sampling techniques for social scientists, industrial managers, and others who find statistics important at work. 61 tables. 90 figures. xvii +602pp. 5⅜ x 8½. 64684-X

LINEAR PROGRAMMING AND ECONOMIC ANALYSIS, Robert Dorfman, Paul A. Samuelson and Robert M. Solow. First comprehensive treatment of linear programming in standard economic analysis. Game theory, modern welfare economics, Leontief input-output, more. 525pp. 5⅜ x 8½. 65491-5

PROBABILITY: An Introduction, Samuel Goldberg. Excellent basic text covers set theory, probability theory for finite sample spaces, binomial theorem, much more. 360 problems. Bibliographies. 322pp. 5⅜ x 8½. 65252-1

GAMES AND DECISIONS: Introduction and Critical Survey, R. Duncan Luce and Howard Raiffa. Superb nontechnical introduction to game theory, primarily applied to social sciences. Utility theory, zero-sum games, n-person games, decision-making, much more. Bibliography. 509pp. 5⅜ x 8½. 65943-7

INTRODUCTION TO THE THEORY OF GAMES, J. C. C. McKinsey. This comprehensive overview of the mathematical theory of games illustrates applications to situations involving conflicts of interest, including economic, social, political, and military contexts. Appropriate for advanced undergraduate and graduate courses; advanced calculus a prerequisite. 1952 ed. x+372pp. 5⅜ x 8½. 42811-7

FIFTY CHALLENGING PROBLEMS IN PROBABILITY WITH SOLUTIONS, Frederick Mosteller. Remarkable puzzlers, graded in difficulty, illustrate elementary and advanced aspects of probability. Detailed solutions. 88pp. 5⅜ x 8½. 65355-2

PROBABILITY THEORY: A Concise Course, Y. A. Rozanov. Highly readable, self-contained introduction covers combination of events, dependent events, Bernoulli trials, etc. 148pp. 5⅜ x 8¼. 63544-9

STATISTICAL METHOD FROM THE VIEWPOINT OF QUALITY CONTROL, Walter A. Shewhart. Important text explains regulation of variables, uses of statistical control to achieve quality control in industry, agriculture, other areas. 192pp. 5⅜ x 8½. 65232-7

Math–Geometry and Topology

ELEMENTARY CONCEPTS OF TOPOLOGY, Paul Alexandroff. Elegant, intuitive approach to topology from set-theoretic topology to Betti groups; how concepts of topology are useful in math and physics. 25 figures. 57pp. 5⅜ x 8½. 60747-X

COMBINATORIAL TOPOLOGY, P. S. Alexandrov. Clearly written, well-organized, three-part text begins by dealing with certain classic problems without using the formal techniques of homology theory and advances to the central concept, the Betti groups. Numerous detailed examples. 654pp. 5⅜ x 8½. 40179-0

EXPERIMENTS IN TOPOLOGY, Stephen Barr. Classic, lively explanation of one of the byways of mathematics. Klein bottles, Moebius strips, projective planes, map coloring, problem of the Koenigsberg bridges, much more, described with clarity and wit. 43 figures. 210pp. 5⅜ x 8½. 25933-1

CONFORMAL MAPPING ON RIEMANN SURFACES, Harvey Cohn. Lucid, insightful book presents ideal coverage of subject. 334 exercises make book perfect for self-study. 55 figures. 352pp. 5⅝ x 8¼. 64025-6

THE GEOMETRY OF RENÉ DESCARTES, René Descartes. The great work founded analytical geometry. Original French text, Descartes's own diagrams, together with definitive Smith-Latham translation. 244pp. 5⅜ x 8½. 60068-8

PRACTICAL CONIC SECTIONS: The Geometric Properties of Ellipses, Parabolas and Hyperbolas, J. W. Downs. This text shows how to create ellipses, parabolas, and hyperbolas. It also presents historical background on their ancient origins and describes the reflective properties and roles of curves in design applications. 1993 ed. 98 figures. xii+100pp. 6½ x 9¼. 42876-1

THE THIRTEEN BOOKS OF EUCLID'S ELEMENTS, translated with introduction and commentary by Thomas L. Heath. Definitive edition. Textual and linguistic notes, mathematical analysis. 2,500 years of critical commentary. Unabridged. 1,414pp. 5⅜ x 8½. Three-vol. set. Vol. I: 60088-2 Vol. II: 60089-0 Vol. III: 60090-4

GEOMETRY OF COMPLEX NUMBERS, Hans Schwerdtfeger. Illuminating, widely praised book on analytic geometry of circles, the Moebius transformation, and two-dimensional non-Euclidean geometries. 200pp. 5⅝ x 8¼. 63830-8

DIFFERENTIAL GEOMETRY, Heinrich W. Guggenheimer. Local differential geometry as an application of advanced calculus and linear algebra. Curvature, transformation groups, surfaces, more. Exercises. 62 figures. 378pp. 5⅜ x 8½. 63433-7

CURVATURE AND HOMOLOGY: Enlarged Edition, Samuel I. Goldberg. Revised edition examines topology of differentiable manifolds; curvature, homology of Riemannian manifolds; compact Lie groups; complex manifolds; curvature, homology of Kaehler manifolds. New Preface. Four new appendixes. 416pp. 5⅜ x 8½. 40207-X

History of Math

THE WORKS OF ARCHIMEDES, Archimedes (T. L. Heath, ed.). Topics include the famous problems of the ratio of the areas of a cylinder and an inscribed sphere; the measurement of a circle; the properties of conoids, spheroids, and spirals; and the quadrature of the parabola. Informative introduction. clxxxvi+326pp; supplement, 52pp. 5⅜ x 8½. 42084-1

A SHORT ACCOUNT OF THE HISTORY OF MATHEMATICS, W. W. Rouse Ball. One of clearest, most authoritative surveys from the Egyptians and Phoenicians through 19th-century figures such as Grassman, Galois, Riemann. Fourth edition. 522pp. 5⅜ x 8½. 20630-0

THE HISTORY OF THE CALCULUS AND ITS CONCEPTUAL DEVELOPMENT, Carl B. Boyer. Origins in antiquity, medieval contributions, work of Newton, Leibniz, rigorous formulation. Treatment is verbal. 346pp. 5⅜ x 8½. 60509-4

THE HISTORICAL ROOTS OF ELEMENTARY MATHEMATICS, Lucas N. H. Bunt, Phillip S. Jones, and Jack D. Bedient. Fundamental underpinnings of modern arithmetic, algebra, geometry, and number systems derived from ancient civilizations. 320pp. 5⅜ x 8½. 25563-8

A HISTORY OF MATHEMATICAL NOTATIONS, Florian Cajori. This classic study notes the first appearance of a mathematical symbol and its origin, the competition it encountered, its spread among writers in different countries, its rise to popularity, its eventual decline or ultimate survival. Original 1929 two-volume edition presented here in one volume. xxviii+820pp. 5⅜ x 8½. 67766-4

GAMES, GODS & GAMBLING: A History of Probability and Statistical Ideas, F. N. David. Episodes from the lives of Galileo, Fermat, Pascal, and others illustrate this fascinating account of the roots of mathematics. Features thought-provoking references to classics, archaeology, biography, poetry. 1962 edition. 304pp. 5⅜ x 8½. (Available in U.S. only.) 40023-9

OF MEN AND NUMBERS: The Story of the Great Mathematicians, Jane Muir. Fascinating accounts of the lives and accomplishments of history's greatest mathematical minds–Pythagoras, Descartes, Euler, Pascal, Cantor, many more. Anecdotal, illuminating. 30 diagrams. Bibliography. 256pp. 5⅜ x 8½. 28973-7

HISTORY OF MATHEMATICS, David E. Smith. Nontechnical survey from ancient Greece and Orient to late 19th century; evolution of arithmetic, geometry, trigonometry, calculating devices, algebra, the calculus. 362 illustrations. 1,355pp. 5⅜ x 8½. Two-vol. set. Vol. I: 20429-4 Vol. II: 20430-8

A CONCISE HISTORY OF MATHEMATICS, Dirk J. Struik. The best brief history of mathematics. Stresses origins and covers every major figure from ancient Near East to 19th century. 41 illustrations. 195pp. 5⅜ x 8½. 60255-9

Physics

OPTICAL RESONANCE AND TWO-LEVEL ATOMS, L. Allen and J. H. Eberly. Clear, comprehensive introduction to basic principles behind all quantum optical resonance phenomena. 53 illustrations. Preface. Index. 256pp. 5⅜ x 8½. 65533-4

QUANTUM THEORY, David Bohm. This advanced undergraduate-level text presents the quantum theory in terms of qualitative and imaginative concepts, followed by specific applications worked out in mathematical detail. Preface. Index. 655pp. 5⅜ x 8½. 65969-0

ATOMIC PHYSICS: 8th edition, Max Born. Nobel laureate's lucid treatment of kinetic theory of gases, elementary particles, nuclear atom, wave-corpuscles, atomic structure and spectral lines, much more. Over 40 appendices, bibliography. 495pp. 5⅜ x 8½. 65984-4

A SOPHISTICATE'S PRIMER OF RELATIVITY, P. W. Bridgman. Geared toward readers already acquainted with special relativity, this book transcends the view of theory as a working tool to answer natural questions: What is a frame of reference? What is a "law of nature"? What is the role of the "observer"? Extensive treatment, written in terms accessible to those without a scientific background. 1983 ed. xlviii+172pp. 5⅜ x 8½. 42549-5

AN INTRODUCTION TO HAMILTONIAN OPTICS, H. A. Buchdahl. Detailed account of the Hamiltonian treatment of aberration theory in geometrical optics. Many classes of optical systems defined in terms of the symmetries they possess. Problems with detailed solutions. 1970 edition. xv+360pp. 5⅜ x 8½. 67597-1

PRIMER OF QUANTUM MECHANICS, Marvin Chester. Introductory text examines the classical quantum bead on a track: its state and representations; operator eigenvalues; harmonic oscillator and bound bead in a symmetric force field; and bead in a spherical shell. Other topics include spin, matrices, and the structure of quantum mechanics; the simplest atom; indistinguishable particles; and stationary-state perturbation theory. 1992 ed. xiv+314pp. 6⅛ x 9¼. 42878-8

LECTURES ON QUANTUM MECHANICS, Paul A. M. Dirac. Four concise, brilliant lectures on mathematical methods in quantum mechanics from Nobel Prize–winning quantum pioneer build on idea of visualizing quantum theory through the use of classical mechanics. 96pp. 5⅜ x 8½. 41713-1

THIRTY YEARS THAT SHOOK PHYSICS: The Story of Quantum Theory, George Gamow. Lucid, accessible introduction to influential theory of energy and matter. Careful explanations of Dirac's anti-particles, Bohr's model of the atom, much more. 12 plates. Numerous drawings. 240pp. 5⅜ x 8½. 24895-X

ELECTRONIC STRUCTURE AND THE PROPERTIES OF SOLIDS: The Physics of the Chemical Bond, Walter A. Harrison. Innovative text offers basic understanding of the electronic structure of covalent and ionic solids, simple metals, transition metals and their compounds. Problems. 1980 edition. 582pp. 6⅛ x 9¼. 66021-4